Understand Electronics

Understand Electronics

Second Edition

Owen Bishop

Newnes

OXFORD AUCKLAND BOSTON JOHANNESBURG MELBOURNE NEW DELHI

Newnes
An imprint of Butterworth-Heinemann Ltd
Linacre House, Jordan Hill, Oxford OX2 8DP
225 Wildwood Avenue, Woburn, MA 01801-2041
A division of Reed Educational and Professional Publishing Ltd

A member of the Reed Elsevier plc group

First published 1995
Reprinted 1996, 1998, 1999
Second Edition 2001
© Owen Bishop 1995
© Owen Bishop 2001

British Library Cataloguing in Publication Data
A catalogue record for this book is available from the British Library

Library of Congress Cataloguing in Publication Data
A catalogue record for this book is available from the Library of Congress

ISBN 0 7506 5391 1

Printed and bound in Great Britain by Biddles Ltd, *www.biddles.co.uk*

PLANT A TREE

BTCV
British Trust for Conservation Volunteers

FOR EVERY TITLE THAT WE PUBLISH, BUTTERWORTH-HEINEMANN
WILL PAY FOR BTCV TO PLANT AND CARE FOR A TREE.

Contents

INTRODUCTION

This is a book for anyone who wants to get to know about electronics. It requires no previous knowledge of the subject, or of electrical theory, and the treatment is entirely non-mathematical. It begins with an outline of electricity and the laws that govern its behaviour in circuits. Then it describes the basic electronic components and how they are used in simple electronic circuits. Semiconductors are given a full treatment since they are at the heart of almost all modern electronic devices. In the next few chapters we examine a range of electronic sensors, seeing how they work and how they are used to put electronic circuits in contact with the world around them.

The methods used for constructing electronic circuits from individual components and the techniques of manufacturing complex integrated circuits on single silicon chips are covered in sufficient detail to allow the reader to understand the steps taken in the production of an item of electronic equipment. This is followed by an account of the test equipment used to check the finished product.

The next few chapters deal with the electronic circuits that are used in special fields and serves as an introduction to amplifiers, logic circuits, audio equipment, computing, telecommunications (including TV and video equipment) and microwave technology. Then we look at the ways in which electronics plays an ever-increasing role in measurement, detection and control in industry and other fields. Throughout, the descriptions are intentionally aimed at the non-technical reader.

Finally, we outline some of the current research in electronics and point the way to future developments in this technology.

All the photographic illustrations in this book were taken by the author.

1 Electrons and electricity

Electricity consists of electric charge. Though electricity has been the subject of scientific investigations for thousands of years, the nature of electric charge is not fully understood, even at the present day. But we do know enough about it to be able to use it in many ways. Using electric charge is what this book is about.

Electric charge is a property of matter and, since matter consists of atoms, we need to look closely at atoms to find out more about electricity. But, even without studying atoms as such, we are easily able to discover some of the properties of electricity for ourselves.

The simplest way to demonstrate electric charge is to take a plastic ruler and rub it with a dry cloth. If you hold the ruler over a table on which there are some small scraps of thin paper or scraps of plastic film, the pieces jump up and down repeatedly. If you rub an inflated rubber balloon against the sleeve of your clothing then place it against the wall or ceiling, for a while, the balloon remains attracted to the wall or ceiling, defying the force of gravity. The electric charge on the plastic ruler or wall is creating a force, an **electric force**. In effect, the energy of your rubbing appears in another form which moves the pieces of paper, or prevents the balloon from falling.

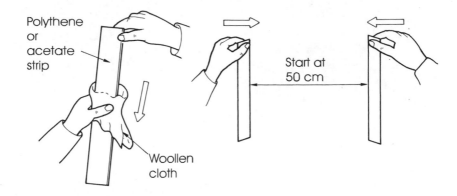

Polythene or acetate strip

Start at 50 cm

Woollen cloth

The nature of electric charge is easy to demonstrate.

You can also charge a strip of polythene sheet (cut from a plastic food-bag) by rubbing it. If you charge two strips, then hold them apart and then try to bring them together, they are repelled by each other. As you try to push them together, their lower ends diverge, spreading away from each other.

A similar experiment is to charge a strip of acetate sheet (cut from a shirt-box) by rubbing it and bring it toward a charged polythene strip, the two strips attract each other. If they are allowed to, their lower ends come together and touch. From this behaviour, we reason that the charge on an acetate strip must be of a different kind from that on a polythene strip. Simple demonstrations such as these show that:

- There are two kinds of electric charge.
- Like kinds of charge repel each other.
- Opposite kinds of charge attract each other.

The discovery of electricity

Electricity takes its name from the Greek word *elektron*, the name of the resinous solid known as amber. The ancient Greeks had discovered that, when a piece of amber is rubbed with a soft cloth, it becomes able to attract small, light objects to it. We say that it has an electric charge.

More electric attraction

You may have noticed this effect in the shower. The fine spray of water droplets charges your body and the shower curtain, but the charges are opposite. There is an attractive force between your body and the curtain. The curtain billows inward and clings to your body.

Electric charge and atoms

Now we are ready to link the basic facts about electric charge to what is known about the structure of matter.

Research has shown that atoms are built up of several different kinds of atomic particle. Most of these occur only rarely in atoms but two kinds are very common. These are electrons and protons. Although protons are about 2000 times more massive than electrons, protons and electrons have equal electric charges. The charge on an electron is opposite in its nature to the charge on a proton; these are the two kinds of electric charge mentioned above. The charge on an electron is said to be negative and that on a proton to be positive, but this is simply a convention. There is nothing positive on a proton which is 'missing' or 'absent' from an electron. The two terms merely imply that positive and negative charges are opposite.

We have said that electrons and protons carry equal but opposite charges. If an electron combines with a proton, their charges cancel out exactly and an uncharged particle is formed — a neutron. Since neutrons have no charge, they are of little interest in electronics.

Atomic structure

All atoms are composed of electrons and protons (ignoring the other rare kinds of particle). The simplest possible atom, the atom of hydrogen, consists of one electron and one proton. The proton is at the centre of the atom and the electron is circling around it in orbit. With one unit of negative charge and one of positive charge, the atom as a whole is uncharged.

Since the electron is moving at high speed around the proton, there must be a force to keep it in orbit, to prevent it from flying off into space. The force that holds the electron in the atom is the attractive electrical force between oppositely charged particles, which we demonstrated earlier. It acts in a similar way to the attractive force of gravity, which keeps the Moon circling round the Earth, and the planets of the solar system circling round the Sun.

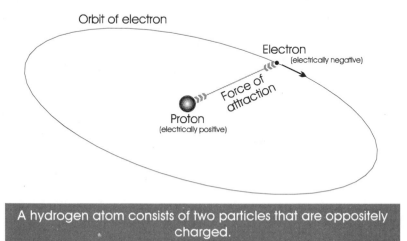

Orbit of electron

Electron
(electrically negative)

Force of attraction

Proton
(electrically positive)

A hydrogen atom consists of two particles that are oppositely charged.

Other atoms

There are more than a hundred different elements in nature, including hydrogen, helium, copper, iron, mercury and oxygen, to name only a few. Each element has its own distinctive atomic structure, but all are based on the same plan as the hydrogen atom. That is to say, there is a central part, the nucleus, where most of the mass is concentrated, which is surrounded by a cloud of circling electrons. However, atoms other than hydrogen have more than one proton and also some neutrons in the nucleus at the centre of the atom. The positive charge on the nucleus is due to the protons it contains. The electron cloud contains a number of electrons to equal the number of protons in the nucleus. In this way the positive charge on the nucleus is exactly balanced by the negative charges on the electrons and the atom as a whole has no electric charge.

The electrons are in orbits at different distances from the nucleus. These orbits are at definite fixed distances from the nucleus and there is room for only a fixed number of electrons in each orbit.

Atomic dimensions

The orbit of the electron of a hydrogen atom is about one ten-millionth of a millimetre in diameter. If the atom was scaled up so that its nucleus (the proton) was 1 mm in diameter, the orbiting electron would be a tiny speck about 120 m away. The interesting point is that the electron and proton take up very little room in the atom. So-called 'solid' matter is mostly empty space. The tangible nature of matter is not due to it consisting mostly of firm particles. Instead, it is due to the strong electrical forces between atomic particles and the forces between adjacent atoms, which hold the atoms more-or-less firmly together. There is more about the structure of matter on page 8.

Electric fields

When an object is charged there is an electric field around it. This is a **force field** which makes charged objects move when they are in the field. Another more familiar force field is gravity, which affects us everywhere and at all times; but gravity is only attractive, it does not repel.

The drawing shows how we imagine the field around an electron. The lines of force show the way a positive charge moves when placed in the field; it moves towards the electron. Although lines of force are strictly imaginary (just as the lines of latitude and longitude on the Earth are imaginary) it helps to think of them as if they are like rubber bands under tension. This gives them two properties:

electron

line of force

- They tend to be as short as possible.
- They tend to be as straight as possible.

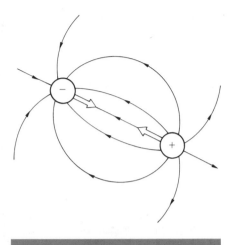

Unlike charges attract

When there is a proton in the vicinity of an electron, we see the fields of the electron and proton combining to give lines running from the proton to the electron. Making the lines as short as possible creates forces acting on the electron and proton, drawing them together until they meet. They are *attracted* to each other.

The field between two electrons is very different. The lines of force of one electron do not join with those of the other electron. Each electron maintains its own field.

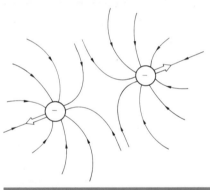

Like charges repel

Another property of lines of force is that they can not cross. So the fields become distorted, as shown. But lines of force tend to become straight and, for this to happen, the electrons are forced to move further apart. They are *repelled* by each other.

For the same reason, two positive charges repel each other.

Charge and energy

We can now begin to understand what happens when we charge a strip of plastic by rubbing it with a cloth. Although there is normally a fixed number of electrons circling around the nucleus of an atom, the electrons in the outer orbits are less strongly attracted to the nucleus than those closer to it. Rubbing the plastic with a cloth provides energy (derived from our muscles) to overcome the attractive forces between the nucleus and the outer electrons. Depending on the nature of the plastic, we may remove electrons from some of the atoms in the molecules of the plastic and attach them to the atoms in the molecules of the cloth.

Removing electrons leaves the plastic with excess positive charge; collecting electrons on the cloth gives it negative charge. When we pull the cloth away from the plastic at the end of the rubbing process, the strip and cloth are oppositely charged, so they are attracted to each other. We need to use more muscular energy to pull them apart now than if they were uncharged.

When the strip and cloth have been separated, they remain charged until charged molecules in the air are attracted to them. The plastic which, lacking electrons, is positively charged, attracts any negatively charged molecules that happen to be in the surrounding air. The electrons on these are passed across to the charged atoms of the plastic, so gradually discharging them. A similar process discharges the cloth.

Charging other substances

Many kinds of substance are charged when rubbed with a cloth: possible substances include different kinds of plastic, rubber, and glass. Exactly what happens depends on which substance is rubbed with which kind of cloth. On this page, we described how the substance becomes negatively charged and the cloth becomes positively charged. But with a different substance or a different kind of cloth, charging may occur in the opposite direction.

Conduction

Substances such as plastic, wool, glass, and rubber can be charged because there is no quick way that electrons removed from an atom can be replaced or that excess electrons can be got rid of. The charged atoms are isolated from each other and from the surroundings, except when, for example, a charged molecule is attracted from the air, or the substance is brought into contact with an oppositely charged surface. Substances of this kind, in which charges stay in a fixed place, are known as **non-conductors** or **insulators**. As well as those mentioned above, this class of substance includes dry wood, paper, ceramics, pure water, asbestos and dry air. The other major class of substance comprises the **conductors,** in which electrons can move freely. These are mostly elements such as silver, gold, copper, lead, and carbon. The majority of them are metals. Conductors also include alloys of metals (such as brass) and solution of salts.

Metals have the structure of crystals, the individual atoms are arranged in a regular three-dimensional array known as a **matrix** or **lattice**. They are held in position by forces (not electric) existing between each atom and its neighbours. These are indicated by the 3D grid of lines in the diagram. Each atom has electrons circling its nucleus but, in metals, the outer electrons are only weakly attracted to the nucleus. They are able to leave the atom and wander off at random into the spaces within the lattice, not being attached to any particular atom. When these electrons move, negative charge moves from one part of the copper block to another. We say that the electrons are **charge carriers**.

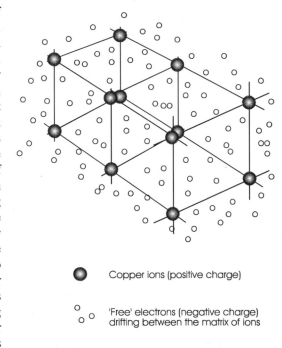

● Copper ions (positive charge)

○ ○
○ ○ 'Free' electrons (negative charge)
       drifting between the matrix of ions

The arrangement of copper atoms in the lattice of metallic copper. A cloud of 'free electrons' permeates the lattice. These electrons are available for conducting charge.

The photograph shows one way in which the electrons may be made to move in an orderly fashion. The filament lamp is connected to an electric battery by copper wires. There is an electric field between the positive and negative terminals of the battery. The way this field is generated is described later but, for the moment, consider there to be lines of force running from the positive terminal (+) to the negative terminal (−). Before the battery is connected to the lamp, the field lines run directly between the terminals, through the air. When the battery is connected by wires to the lamp, the field lines mostly become bunched together, and run through the wires and the filament of the lamp, instead of running through the air. Thus there is an electric field running from the positive terminal, through the wire on the left, through the lamp, through the wire on the right and back to the battery at its negative terminal. Any charged particles in that field will move, provided that they are free to do so.

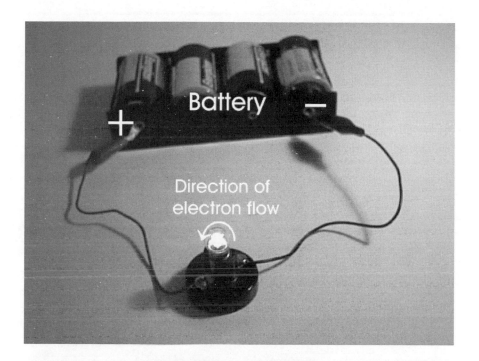

Battery

+   —

Direction of
electron flow

This is a simple example of an electric circuit. Current flows from the
battery, through the various conductors and returns to the battery.

If the wires and filament of the lamp were made of plastic or some other
non-conductor, charged particles would not be free to move. However, in a
conductor such as copper the electrons wandering between the atoms are very
free to move. As long as there is no field within the conductors the electrons
wander randomly in the space within the lattice. Once a field has been applied to
the conductors the electrons all flow in one direction. They flow through the
wires and lamp repelled by the negative terminal of the battery and attracted
toward its positive terminal. We have an **electric current**. This is just what an
electric current is – a mass flow of electric charge from one place to another.

As electrons move through the circuit, those reaching the positive terminal of the
battery pass into it. They are replaced by electrons coming from the negative
terminal. Most of electronics deals with the flow of electric currents. We are not
much interested in the stationary (or static) charges built up by rubbing non-
conductors, but there is one instance in which static charges are really important
and we must take special precautions to eliminate them, as explained on page 83.

Electromotive force

When a charged particle is in an electric field it is subject to a force, which makes it move if it is free to do so. This force is known as electromotive force, often referred to briefly as e.m.f.

Current through a solution

Electrons are not the only particles that can carry an electric charge. The charge carriers in a solution in water are the **ions** of the dissolved substance. For example, when copper sulphate ($CuSO_4$) is dissolved in water, its molecules break up into two ions: copper (Cu) and sulphate (SO_4). When they break up, or **ionize**, the sulphate ion takes two electrons from the copper atom, leaving it positively charged (Cu^{++}). This makes the sulphate ion negatively charged (SO_4^{--}).

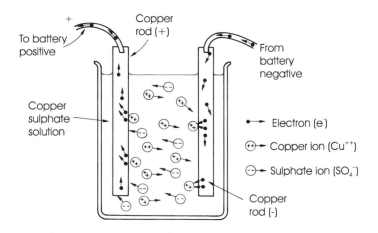

There is a two-way flow of charge through a solution.

Consider the flow of charge carriers between two copper rods immersed in a solution of copper sulphate and connected externally to a battery. The copper ions are attracted toward the negative rod and are there discharged by electrons which have come from the negative terminal of the battery. The discharged copper ions are deposited on the rod as a bright reddish layer of copper.

Copper *atoms* of the positive rod dissolve in the water, becoming copper *ions*, each losing two electrons. The electrons flow to the positive terminal of the battery. This copper rod gradually becomes thinner. The sulphate ions are attracted toward the positive supply but do not become discharged, so they are not charge carriers.

Although salts form ions when dissolved in water, making the salt solution a conductor, pure water does not form ions, and so it is a non-conductor.

Current through a gas

Gases under low pressure can conduct electric charge. As in the case of a solution, there is two-way conduction.

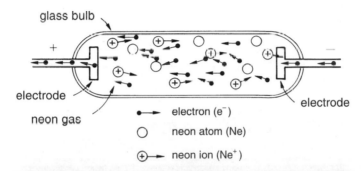

Neon, argon, krypton, and some other gases conduct electric charge when they are at low pressure.

Electrons flow from the negative plate (negative electrode) to the positive plate (positive electrode). On their way, they strike neon atoms and knock electrons out of them. This creates more electrons to act as negative charge carriers. The neon atoms which have lost electrons become positive ions, and act as positive charge carriers. The energy from the moving carriers excites many of the neon atoms. Excited atoms later lose this energy, which then appears in another form, that of reddish light. An example is the 'flicker flame' lamp shown in the title photograph of this chapter.

T he terms **electromotive force** and **potential** are often used as if they mean the same thing. Although they have some features in common, they are not quite identical. This chapter helps you to understand the difference between them.

A source of electromotive force

Electric cells are very often used as sources of electromotive force (e.m.f.). In Chapter 1, we explained how the e.m.f. of the cell gives rise to a field between the terminals of the cell, and how the force produced by this field drives electrons around the circuit from the negative terminal to the positive terminal of the cell.

There are many types of cell but, to illustrate the principles of the way cells work, we will look at a simple **wet cell**. One of the simplest ways of making a wet cell is to take an iron nail and a copper nail (or thick wires) and press them into a lemon. They must not touch each other. You will be able to measure a small voltage between the two nails. This experiment illustrates the main features of cells in general:

- **electrodes** – two, made of different materials. In the cell shown opposite, they are made of copper and zinc.
- **electrolyte** – a conductive fluid or paste. Conduction is by means of ions produced by dissolved substances. In the 'lemon' cell, the electrolyte is the juicy flesh of the lemon.

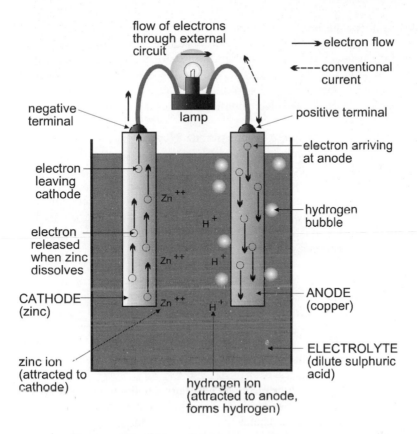

flow of electrons
through external
circuit

⟶ electron flow

⟵--- conventional
current

negative
terminal

lamp

positive terminal

electron
leaving
cathode

electron arriving
at anode

Zn^{++}

H^+

hydrogen
bubble

electron
released
when zinc
dissolves

Zn^{++} H^+

CATHODE
(zinc)

Zn^{++} H^+

ANODE
(copper)

ELECTROLYTE
(dilute sulphuric
acid)

zinc ion
(attracted to
cathode)

hydrogen ion
(attracted to anode,
forms hydrogen)

A simple wet cell produces an e.m.f.

Except for the lead-acid storage cell (or accumulator, see later) wet cells are very rarely used today. For reasons of convenience and safety, they have been superseded by dry cells of many kinds. However, the action of a wet cell clearly illustrates the way cells work.

The copper and zinc electrodes of a wet cell are immersed in dilute sulphuric acid. This is ionized (like the copper sulphate solution described on p. 10) into hydrogen (H^+) and sulphate (SO_4^{--}) ions. Some of the zinc atoms of the cathode dissolve in the acid, forming zinc ions (Zn^{++}). The electrons freed when the zinc dissolves are left in the zinc electrode, giving it a negative charge. The action is driven by chemical energy released as a result of the zinc going into solution. It continues for a while until the increasing negative charge on the zinc electrode attracts zinc ions back to the electrode in such quantities that no further dissolving occurs.

At this point, we are left with the zinc electrode being negatively charged with respect to the electrolyte and to the copper electrode. There is an electric field between the cathode and anode. If the terminals of the cell are connected by wires to a lamp or other conductive device the electric field between the electrodes *forces* electrons to *move* from the zinc cathode, through the wires, to the copper anode. This is the **electromotive force** of the cell.

The cell begins to supply an **electric current**. Now that electrons are flowing out of the cathode and into the connecting wire, the cathode becomes less negatively charged than before and attracts zinc ions less strongly. This means that more zinc is able to dissolve, producing a further supply of electrons. Electrons that have flowed through the wires reach the anode. There they attract H^+ ions from the electrolyte. Being negatively charged, the electrons discharge the hydrogen ions, producing uncharged hydrogen atoms that collect together to form bubbles of hydrogen gas.

As long as there is an electrical connection between the terminals of the cell, the action of the cell continues as described above. The e.m.f. forces a current of electrons to flow from the cathode to the anode. The cathode is steadily eaten away as the zinc dissolves (this is what is driving the action, converting chemical energy into electrical energy) and bubbles of hydrogen gas rise steadily from the anode. The process continues until the zinc is completely dissolved.

Other sources of e.m.f. include electrical generators, deriving their energy from coal, oil, atomic power, water turbines, the wind, or the waves. Solar panels, such as are used on satellites, generate e.m.f. from the energy of sunlight.

Which way does the current flow?

The drawing of the wet cell shows a flow of electrons from the negative terminal of the cell to the positive terminal. Yet we are accustomed to thinking of current as flowing from positive to negative. Most of the diagrams in this book show it flowing in that direction. Current flowing from positive to negative is known as **conventional current**. It is just a convention that this is the direction in which it flows. But, when we look more closely at the charge carriers themselves, we often find that the *actual* flow is that of negative charge carriers (electrons) from negative to positive.

Other cells

The wet cell is seldom used because it possesses several disadvantages. The acid electrolyte is a corrosive substance and makes the cell hazardous to handle. The fact that it generates hydrogen gas means that the cell can not be sealed to contain the acid safely. Hydrogen gas is explosive, so this is another source of danger. Other types of cell are available which work on the same principles, converting chemical energy to electrical energy, but with greater safety and with the ability to produce larger current.

Most other types of cell are **dry cells**, in which the liquid electrolyte is replaced by a stiff paste. In all types of cell the electrodes are of different materials. One of the electrodes may be shaped to become the container of the cell. For example, in the popular zinc-carbon cell (the sort we often use in an electric torch), the cylindrical container is made of zinc and is the cathode. The anode is a carbon rod running down the axis of the cell.

A variety of dry cells is available, having different electrical features:

Types of dry cell	
Zinc-carbon	The typical cheap 'torch cell', liable to leak when old.
Alkaline	Can supply a large current. Holds about three times as much charge as a zinc-carbon cell. Has low leakage. Often used in torches, radios, and recorders.
Zinc chloride	Can supply a large current. Holds more charge than a zinc-carbon cell, and has low leakage.
Silver oxide	Low power but long lasting. Made as button cells for watches and calculators.
Mercury oxide	Similar to silver oxide. Used in hearing aids.

Types of dry cell (continued)	
Lithium	There are three variations: Li-manganese, Li-iron disulphide and Li-thionyl chloride. Deliver low currents for a long time (years) or a high current (up to 30 A) for a relatively short time. Store about three times as much charge as alkaline cells, and remain in working condition for many (up to 10) years. Expensive. They are often used as back-up supply for computer memory. Research is producing flexible Li cells,cased in plastic, that can be bent to fit into confined or oddly-shaped spaces in portable equipment.
Zinc air	High power in small size. Used for laptop computers.

Cells and batteries

A cell is a single unit. A battery consists of several cells connected together. Usually the cells of a battery are joined cathode-to-anode. The advantage of joining cells in this way is that the e.m.f. of the battery equals the total of the e.m.f.s of the individual cells.

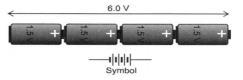

The electric field is that much stronger and so is the current produced.

The wet cells and dry cells descibed so far are known as **primary cells**. Their e.m.f. comes from the nature of the chemicals used in making them. The other main category of cells is the **secondary cell** or **rechargeable cell**. The materials from which the cell is assembled do not produce any e.m.f. When the cell is **charged** by applying an e.m.f. from an external source, some of the materials in the cell change their chemical composition and produce an e.m.f.

Types of rechargeable cell	
Nickel-cadmium (NiCad)	Store less charge than the zinc-carbon cells, but can deliver a high current. One disadvantage is that, if they are not allowed to discharge fully before recharging, they `remember' the level at which they were recharged. Subsequently, they cease to deliver current when they have been discharged to the `remembered' level, reducing their capacity appreciably.
Lead-acid (Accumulator)	Very high current, high charge. Used as car batteries and for similar purposes. Heavy.
Nickel metal-hydride (NiMH)	Have several advantages over the NiCad type, one of which is that they are able to store 30–50% more energy in a given volume. They do not exhibit the `memory effect' so it is not necessary to discharge them completely before re-charging. They are particularly suited to use in electrically-powered vehicles and video cameras.

Fuel cells

A fuel cell is a device that produces e.m.f from a continuous supply of energy-rich fuels. It will continue to produce the e.m.f., and therefore an electric current, for as long as we supply it with fuel. In some ways it is like a petrol engine. It keeps running and produces useful energy for as long as its fuel tank is kept topped up. Unlike a petrol engine, the waste product from the cell is environmentally harmless. In many cases it is just water.

The idea of fuel cells was first put forward in the early nineteenth century but there were many practical problems to solve. In the late twentieth century the Space Shuttle Program provided the spur to increased research on fuel cells. This successfully resulted in the use of fuel cells to provide most of the electrical power on these vehicles. With the increasing use of portable equipment such as lap-top computers and cellphones, development of fuel cells is proceeding at an even greater rate.

A fuel cell has much in common with the types of cell that we have already described. There are two electrodes (cathode and anode) immersed in a medium containing chemically active substances. The energy released when a chemical reaction occurs is converted to an e.m.f. In the wet cell the reaction is the dissolving of zinc in acid to form zinc sulphate solution. In a fuel cell, the e.m.f. is usually the result of the combination of hydrogen and oxygen to form water. Under the right circumstances, hydrogen and oxygen combine with a bang, releasing large amounts of energy, but the process is under strict control in the fuel cell.

There are several types of fuel cell. The **proton-exchange membrane cell** (PEM) has two electrodes coated with platinum to act as a catalyst. Hydrogen is supplied as gaseous hydrogen or more often in the form of a hydrogen-rich substance such as methanol. In the vicinity of the cathode the hydrogen ionizes to an electron and a proton (H^+, the nucleus of the hydrogen atom). Electrons accumulate on the cathode and the protons diffuse through the polymer membrane that separates the cathode from the anode. Oxygen (from the air) is available on the other side of the membrane in the vicinity of the anode. Here the atoms of oxygen combine with protons and with electrons from the anode to form atoms of water.

There is a surplus of electrons at the cathode. Electrons are required at the anode. If there is an external circuit connected to the electrodes, electrons flow from cathode to anode. The e.m.f. that drives this flow is powered by the chemical action of the oxygen with the protons and electrons.

The **solid oxide fuel cell** (SOFC), too, relies on the reaction between hydrogen and oxygen to form water. In the SOFC it is the oxygen that is ionized. To ionize oxygen requires a high temperature in the region of 1000°C, so this type of cell is more suited to industrial applications.

An **alkaline fuel cell** consists of two porous electrodes immersed in an electrolyte containing potassium hydroxide. It is said to be cheaper to manufacture than either of the other types of fuel cell, though it is still in the development stage. It does not require a high temperature to operate it.

New cells

Fuel cells may still be ranked among the newer types of cell, though practical versions are fast coming on the market. We may see further developments within a few years. With the increasing use of portable electronic equipment, such as cellphones and lap-top computers, there is an incentive to develop power sources suited to their special needs. The usual aim is to combine high capacity with small physical size. Other desirable features are a constant voltage up to the time of complete discharge and the possibility of recharging. **Polymer batteries** have their anode made from foil of a conductive polymer called polypyrrol. The battery has the desirable properties of lithium batteries. For lap-tops, further work is being done on zinc-air batteries which are able to store a large amount of power in small size. Such batteries will be able to power a lap-top computer for 10 to 20 hours before needing recharging.

Perhaps the most interesting development is the **smart battery.** This has a small integrated circuit (p. 159) in it which links up with a microprocessor in the smart charger. It should be explained that for most effective battery operation and maximum battery life, charging needs to be performed according to a carefully controlled routine. The charging current must be adjusted continually according to the amount of charge already in the battery and the temperature of the battery. Sensors in the battery report on battery status to the microprocessor in the charger and the optimum charging program is selected automatically.

Measuring current, charge and e.m.f.

The fundamental electrical unit is the unit of current, the **ampere**. This is often known as an 'amp' for short, and its symbol is A. The ampere is measured by measuring the force between two coils each carrying the same current.

We do not need to be concerned with the size of the force or the dimensions of the coils. It is enough to know that current can be defined in purely physical terms. Having defined the ampere, we define the unit of charge, the **coulomb** (symbol, C) by saying that if a current of 1 A flows for 1 second, the amount of charge carried is 1 C. A coulomb is a relatively large amount of charge when compared with the charge on an electron. The charge on an electron is only 1.6 $\times 10^{-19}$ C (one sixteen trillionths of a coulomb). This indicates that there must be an unimaginably large number of electrons flowing through the element of an electric kettle when the current through it is, say, 2.5 A.

Another important unit is the unit of **electrical potential**. This is the unit used for expressing e.m.f. To understand this we will look first at a similar idea, the idea of gravitational potential. If we lift a brick up from the ground, we are acting against gravity. We are doing work, or expending energy to lift the brick. If we let the brick go, it drops down again; our energy reappears as the motion of the brick, the noise made as it hits the ground, and possibly the energy expended as it shatters. While we are holding the brick still, above ground, it does not appear to have any extra energy. But it actually has **potential energy**, the energy stored in it because of its *position* above ground. The amount of potential energy (p.e.) stored in the brick depends on two things:

- The mass of the brick.
- The height above ground to which we lift it.

The higher we lift the brick, the greater its potential energy. The greater its potential energy, the more noise it will make when it hits the ground, the more likely it is to dent the ground, and the more likely it is to shatter on impact. If we take several identical bricks and lift them to different heights, the different heights are associated with different amounts of potential energy (per brick). Each point above the ground has a different potential, beginning with zero potential at ground level, and increasing with the distance above ground. We could say that every point has a potential depending on how much energy we need to lift a brick to that point.

The same kind of thinking applies to an electric field. Consider a positively charged object A (opposite), fixed in space with an electric field around it (not shown). There is another positively charged object B, which is a long way from A on the right of the diagram. The force of repulsion between two charged particles depends on the distance between them. It increases as B gets nearer to A. Repulsion is small at first but increases as B approaches A. We are using energy move B because we are acting *against* the force. If we let B go free, it travels back until it is a long way from A. What happens is that the *potential* energy 'stored' in B is recovered as the energy of motion of B away from A.

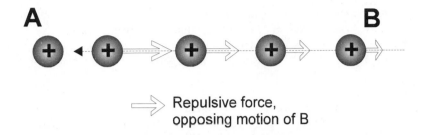

⟹ Repulsive force,
opposing motion of B

Moving a charged particle B up close to A requires work to
overcome the repulsive force between them.

If we bring B close to A and hold it there, it does not appear to have any energy,
but it has potential energy. We can say that:

*The electrical potential in the region of A is measured by the amount of energy
needed to move B to point X from a great distance.*

It makes a better definition if we say that:

*The electrical potential at any point in space is the amount of energy
(measured in joules) needed to move a unit charge (1 C) to that point from a
great distance.*

The unit of electrical potential is *joules per coulomb* (symbol J/C), in other
words, 'How many joules are needed to move a coulomb of charge to that point
in space?' We use the idea of potential so often that we give the unit 'joules per
coulomb' a special name of its own, the **volt**, symbol V.

Potential difference

It is not really a practical task to measure the potential at a point, for example,
the potential at the top left-hand corner of this page in the book. We would have
to start at some point (if it exists) in outer space, far from any stars or galaxies,
and measure how much work we did in bringing a unit charge (if we had the
means of handling it) from there to the book. It is much more important to be
able to measure the *difference* between the potentials at two relatively nearby
points, such as two points in the same electronic circuit. We refer to this as
potential difference. The term 'potential difference' is usually shortened to **p.d.**

The idea of potential difference can be explained by comparing it with potential due to gravity. When we climb a hill, we may like to know how far up it is from bottom to top. We are only interested in the difference in heights, not the actual heights of bottom and top as measured from sea level or from the centre of the Earth.

Consider the electric field between the anode and cathode of a cell. If we place a unit positive charge (1C) near the cathode and move it toward the anode, we have to use energy to overcome the attraction toward the cathode and the repulsion from the anode. Moving the charge toward the anode gives it more potential energy. Its potential is greater there than when it is beside the cathode. There is a potential difference between the two points. For a typical dry cell, the p.d.is about 1.5 V. That is to say, we need to expend 1.5 J of energy to move a 1 C positive charge from the cathode to the anode. Conversely, if a 1 C positive charge is placed at the anode and is allowed to flow (through a wire, for example) from anode to cathode, 1.5 J of energy is released. What becomes of this energy we will see later.

Units, multiples and sub-multiples

The electrical units have a system of multiples and sub-multiples. In electronics we must often use sub-multiples for current and potential:

1 A   = 1000 mA (milliamps)   1V = 1000 mV (millivolts)
1 mA   = 1000 μA (microamps) 1mV = 1000 μV (microvolts)

For the distribution of the electric mains we also use multiples of potential:

1000 V   =1 kV (kilovolts)

Potential and voltage

The electrical potential at a point or the p.d. between two points is measured in volts. 'Potential' and 'potential difference' or 'p.d.' are the strictly correct terms, and we use them as often as we can in this book.

The terms 'voltage' and 'voltage difference', which mean the same thing, are used in a rather more practical sense, particularly in the context of electrical (as opposed to electronic) descriptions. There are instances where these terms seem to sound better than 'potential' and 'p.d.' and then we use them.

E.m.f. and potential

To sum up the main points of this chapter, e.m.f. is a force due to an electric field produced by some kind of generator, such as a cell. The e.m.f. of a cell is the result of the chemical activities taking place inside it. Without a practical generator of some kind, there can be no e.m.f. Potential is a property of each and every point in space, depending how much energy is required to bring unit charge to that point. It is a much more theoretical concept, though potential difference may sometimes be the result of an e.m.f. Both are measured in volts.

For more insight into e.m.f. and potential, see the next chapter.

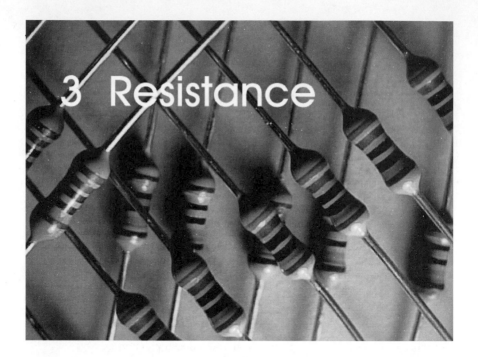

3 Resistance

W hen electrons or other charge carriers flow through a conductor, they move through the lattice and pass close to the atoms of the conductor. They may actually collide with them or at least lose energy to them when they come within range of the electrical fields around the atoms or their nuclei. Whatever happens, the electrons lose energy. We say that the conductor offers **resistance** to the flow of current.

At the same time as the electrons lose energy, the atoms of the conductor gain energy. The effect of gaining energy is usually to make the atoms vibrate slightly about their usual fixed positions in the lattice. The motion of one atom is transferred to other atoms in the lattice, so that all of the atoms are in slight motion. This type of motion is a form of thermal energy, usually known as **heat**. When a current passes through a conductor, the conductor becomes warm or even very hot. In the case of the filament of an electric lamp, it becomes so hot that it glows with visible light. Some of the electrical energy has been converted to heat and some to light energy.

When a charge carrier begins its journey through a conductor it has potential energy, depending on the electrical potential at the point it starts from. During its passage though the conductor it loses some of this potential energy, which is used to heat the conductor. The loss of energy gives rise to a potential difference (p. 21) between one end of the conductor and the other. The amount of energy lost is proportional to the number of charge carriers passing through the conductor, in other words, to the current. We can say that:

The potential difference between two ends of a conductor of a given material is proportional to the current.

This statement is a version of **Ohm's Law**, first stated by Georg Ohm. This law can be put into mathematical form. If a current I is passing through a conductor and there is a p.d. V between its ends, we can calculate R, where:

$$R = \frac{V}{I}$$

We find that, for any given conductor, R is constant. R is called the **resistance** of the conductor. If I is expressed in amps and V in volts, the resistance is in ohms.

For example, a conductor is connected to a battery which has a p.d. of 12 V between its terminals. It is found that a current of 4 A is passing through the conductor. What is its resistance? In this example, $V = 12$ and $I = 4$ so:

$$R = \frac{V}{I} = \frac{12}{4} = 3$$

The resistance is 3 ohms. Instruments for measuring p.d.s and currents are described in Chapter 14.

The **ohm** is the unit of resistance and its symbol is Ω. It has as multiples, the kilohm (1 kΩ = 1000 Ω) and the megohm (1 MΩ = 1000 kΩ).

The resistance of any given conductor also depends on the nature of the conductor. It depends on the size and shape of the conductor and the material from which it is made. Some conducting materials have a bigger supply of charge carriers than others, which makes it easier for them to conduct a current. Examples of good conductors are copper and aluminium.

V and V

We use the symbol V, in upright (Roman) capitals, for the unit of p.d., the volt.

We use the symbols V and v, in slanting (italic) capitals, for the value of any specified p.d.

Resistors

All conductors except superconductors (see p. 33) offer resistance to electric current. In electronics we use special components, called **resistors**, to provide a given amount of resistance as required in a circuit. Various types of resistor are available, one widely used type being the metal film resistor shown in the title photograph of this chapter. This consists of a rod of insulating ceramic coated with a metal film. The film forms a spiral track running from one end of the resistor to the other. It makes contact with the terminal wires at each end. The resistance of the resistor depends upon the length and the width of the track, and upon the metal from which it is made.

Resistors can be made to a very high degree of precision, but precision resistors are unnecessary for many electronic circuits. Resistors are manufactured with a quoted degree of precision, known as **tolerance**. If a resistor has a tolerance of 5%, for example, we know that the actual resistance is within 5% of the nominal value (the value marked on the resistor). For example, if a resistor is marked with a nominal value 33 Ω and its tolerance is indicated as 5%, it means that the actual value of the resistor may be any value within 5% of 33 Ω. Since 5% of 33 is 1.65, the value is between:

$$33 - 1.65 = 31.35 \ \Omega$$
$$\text{and}$$
$$33 + 1.65 = 34.65 \ \Omega$$

If we regard a nominally 33 Ω resistor with 5% tolerance as being close enough for circuit-building, there is no point in making resistors with nominal values in the range 31.35 Ω to 34.65 Ω. The next useful lower value is 30 Ω, for resistors with this nominal value may lie between 28.5 Ω and 31.5 Ω. The range of 30 Ω resistors just touches the range of 33 Ω resistors. Similarly, the next useful value above 33 Ω is 36 Ω, with a range from 34.2 Ω to 37.8 Ω.

To minimize wasteful overlapping of actual resistor values, resistors are made in a series of **preferred values**, known as the **E24 series**. In this series, there are twenty-four basic values:

10, 11, 12, 13, 15, 16, 18, 20, 22, 24, 27, 30,

33, 36, 39, 43, 47, 51, 56, 62, 68, 75, 82, and 91

The series continues with twenty-four values ten times the above: 100, 110, 120, 130, 150, ... , 910.

And then it continues with twenty-four more values: 1000, 1100, 1200, 1300, 1500, ..., 9100 (or 1k, 1.1k, 1.2k, 1.3k, 1.5k, ... , 9.1k).

This scheme is repeated decade by decade up to the maximum practicable value, which is usually 10 MΩ.

There is also a set of smaller values: 1.0, 1.1, 1.2, 1.3, 1.5, ... , 9.1.

In this way the E24 series comprises resistors with 5% tolerance to completely cover the range 1 Ω to 10 MΩ.

Certain types of resistor are made in a restricted range of values, known as the E12 series. In each decade this comprises alternate values of the E24 series: 10, 12, 15, ..., 82. The resistors which are produced as an E12 series have tolerance of 10%, so there is no point in including the other values of the E24 series. The series repeats in each decade as with the E24 series.

For even greater precision, though at greater expense, resistors are made with 2% and 1% tolerance in the E48 and E96 series.

Colour codes

The resistance of a resistor is indicated by a number of coloured bands, the **colour code**, painted on the resistor. Two main systems are used, the four-band system and the five-band system. The four-band system uses the first three bands to indicate resistance. The colours represent numbers according to the following code:

Colour	Number
Black	0
Brown	1
Red	2
Orange	3
Yellow	4
Green	5
Blue	6
Violet	7
Grey	8
White	9

The first two bands (A and B) indicate the first two digits of the value. The third band (C) indicates a multiplier, specified as a power of 10. For example, if the resistance is 6800 Ω, the first two bands are blue and grey.

To obtain the actual resistance, this value must be multiplied by 100, or 10^2. The power of 10 required is 2 and red is the colour code for 2, so the three bands are:

blue, grey, red

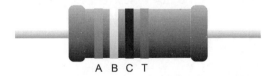

A B C T

In the four-band system, bands A, B, and C indicate the resistance, and band T indicates tolerance.

Another example: if a resistor is marked with yellow, violet, and yellow bands (in that order), the first two bands indicate 47. The yellow third band shows that the multiplier is 10^4, which is 10000. The value is $47 \times 10\,000 = 470\,000$ Ω, which is equivalent to 470 kΩ. This system may seem complicated at first but, in the E24 system, there are only twenty-four possible pairs of colours for the two digits (brown/black, brown/brown, brown/red , ... , white/brown) and these are soon learnt.

For resistors of less than 10 Ω, we use two sub-multipliers:

silver means $\times 0.1$ and *gold* means $\times 0.01$

For example, if the bands are green, blue, gold, the resistance is:

$$56 \times 0.01 = 0.56 \ \Omega.$$

A B C D T

A high-precision resistor is marked with five coloured bands.

The five-band colour code is used for the marking of high-precision resistors in the E48 and E96 series, though it is often used for the E12 and E24 series as well.

The first three bands (A, B, C) indicate the first three digits and the multiplier is indicated by the fourth band (D). Band T shows the tolerance.

Example : The value bands on a 0.5% high precision resistor, value 2.49 kΩ, are red, yellow, white and brown.

The colour codes for tolerance (band T in the figures) are listed in the table.

Colour	Tolerance (±%)
No band	20
Silver	10
Gold	5
Red	2
Brown	1
Green	0.5
Blue	0.25
Violet	0.1

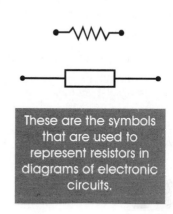

These are the symbols that are used to represent resistors in diagrams of electronic circuits.

Surface mount components

With the increasing sophistication and complexity of circuits on the one hand and the demand for compact, portable equipment on the other hand, any technique that reduces the physical size of components and therefore that of the circuit board, is welcome. An important development in this direction is **surface mount technology**, known as **SMT** for short. This is widely used nowadays in products that need to be small and portable such as pocket telephones and lap-top computers. It is also used for larger equipment in which size reduction gives benefits, for example in personal computers, to reduce the amount of desk space occupied.

SMD resistors and capacitors, type 0603, with an ordinary resistor for comparison .

Surface mount devices (or **SMDs**) do not have wire leads. Instead, they are soldered directly to the board on the same side as the component. Since no holes are needed, one exacting step in circuit board production is completely eliminated, with valuable savings in production costs. The other key feature of SMDs is that they are smaller than the equivalent conventional components.

The dimensions of the commonly used 1206 resistors are 3.2 mm × 1.6 mm. Smaller types include 0805 (2.0 mm × 1.25 mm), 0603 (1.6 mm × 0.8 mm, photograph above), and 0402 (1.0 mm × 0.5 mm). All the conventional types of resistor are available as SMT devices, including power resistors and variable resistors. Fixed resistors are available in all the values of the E96 series.

SMT resistors are too small to be marked with the bands of the colour code. Instead they are usually marked with a 3-digit code. The first two digits represent the first two digits of the resistance, in ohms. The third digit represents the multiplier, as a power of 10. Or this can be thought of as the number of zeros following the two value digits.

Examples: The marking '822' indicates 82×10^2, which is 8200 Ω or 8.2 kΩ. Similarly, the marking '430' indicates 43 Ω.

However, most SMDs are sold on reels of tape for use in automatic assembly machines, and the value is marked on the reel, but not on the device itself.

Connected resistances

Two or more resistances may be connected in **series** or in **parallel.** When they are connected in series, we find their total resistance by simply adding them together.

Example: If R_1 is 47 Ω and R_2 is 82 Ω, the resistance of the two in series is 47 + 82 = 129 Ω.

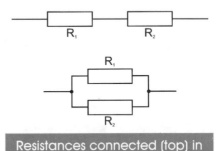

The same rule applies to three or more resistances, all in series.

When two resistances are connected in parallel, we find their total resistance by using this formula:

Resistances connected (top) in series, and (bottom) in parallel.

$$\frac{1}{R} = \frac{1}{R_1} + \frac{1}{R_2}$$

The formula can be extended by adding $1/R_3$ and even more terms to cater for three or more resistances in parallel.

If there are only two resistors in parallel, the formula is simplified to:

$$R = \frac{R_1 \times R_2}{R_1 + R_2}$$

Example: If R_1 is 47 Ω and R_2 is 82 Ω, the resistance of the two in parallel is:

$$R = \frac{47 \times 82}{47 + 82} = \frac{3854}{129} = 29.9\Omega$$

Note that the resistance of two or more parallel resistances is always *less* than that of the *smallest* resistance.

Resistors as p.d. producers

When a current passes through a resistor the charge carriers (electrons) lose potential energy (p. 24). This creates a potential difference between the two ends of the resistor. The size of the p.d. is given by the Ohm's Law equation on page 25. For this reason, we can think of a resistor as a 'current to voltage' converter.

Variable resistors

The resistors described above each have a fixed value. We call them **fixed resistors**. Quite often a circuit needs a resistor which can be varied in value. A typical example is when we want to be able to control the loudness of the sound coming from a radio set.

Variable resistors have a track made of resistive material (such as carbon) with a springy metal wiper that is in contact with the track and is moved along it to vary resistance. The support for the wiper is not shown in the figure.

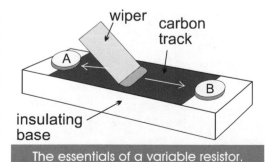

The essentials of a variable resistor.

The resistance between terminal A and the wiper increases as the wiper is moved to the right. At the same time, the resistance between the wiper and terminal B decreases.

Slider resistors or **slider potentiometers**, are very similar in appearance to the figure on page 31, except that they are generally enclosed in a case. They are often used as volume controls in audio systems. Five (possibly more) sliders mounted side-by-side allows the response to be adjusted independently for treble, middle and base frequencies. A **multiturn potentiometer** is similar but the wiper is moved by a threaded rod. It takes many turns of the rod to move the wiper from one end of the track to the other. The result is that the position of the wiper can be very finely adjusted. This feature is useful in circuits that need to be very precisely balanced.

In a **rotary potentiometer** (right) the track is a 270° sector of a circle and the wiper is mounted on the spindle. The whole is enclosed in a metal or plastic case with three terminal pins, so the track is not visible. These potentiometers, or 'pots' as they are often called, have a wide range of uses for controlling electronic circuits. Volume controls, tone controls, brightness controls, and speed controls, may all be effected by a pot.

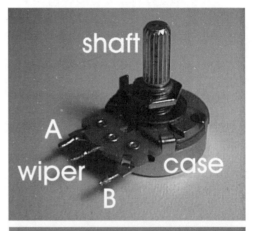

Turning the shaft clockwise increases the resistance between A and the wiper terminal. At the same time, the resistance between B and the wiper terminal is decreased.

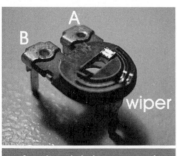

A sub-miniature preset measures only 5 mm between A and B.

A smaller version is the **preset potentiometer**, also known as **trimpot**, which may be enclosed in a case or may be open. Its rotary wiper has a slot to take a screwdriver. Trimpots, as their name implies, are intended to be adjusted when a circuit is being set up, to trim the resistance to the best value once and for all.

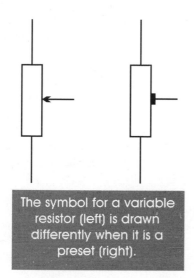

The symbol for a variable resistor (left) is drawn differently when it is a preset (right).

For greater stability, better linearity and longer life, the track of a variable resistor may be made of a metallized ceramic (cermet) instead of a carbon-based material.

Power

As mentioned earlier, charge carriers lose potential when a current flows through a resistance. The energy is transferred to the material of the conductor, making it warm, possibly hot or even white hot.

In an electric motor, most of the energy is used to make the motor rotate, though a little of it appears as heat. In a TV set, the energy is used to produce light and sound, but some heat is produced too. In fact, heat is nearly always produced and usually represents a waste of energy.

The rate at which electrical energy is being used by a device can be calculated very easily, simply by multiplying the current by the p.d. across the device. The result is the **power** of the device, the rate at which it is converting energy from one form (electrical energy) into another form (heat, light, sound, motion). The unit of power measurement is the **watt,** symbol W.

Example: The current through a 6 V torch bulb is found to be 0.5 A. The power of the bulb is $P = 6 \times 0.5 = 3$ W.

This calculation can also be applied to devices which generate current. For example, a solar panel produces a current of 0.8A and the p.d. across its terminals is 12V. Its power is $P = 0.8 \times 12 = 9.6$ W.

Superconduction

Certain materials, called **superconductors**, have the ability to conduct electricity without offering any resistance. This phenomenon was first discovered in 1911 by the Dutch physicist, Heike Kemerlingh Onnes.

The physicist used a wire of solid mercury and found that it had no resistance when the temperature was reduced below 4 K (−269°C). He passed current into a ring of mercury wire cooled in a liquid helium bath and found that the current was still circulating one year later. The temperature below which a material becomes superconducting is known as the **critical temperature, T_C**. Several other substances have since been found to be superconductive, including tungsten, aluminium and lead, but only at critical temperatures of a few kelvin. These are known as Type 1 superconductors. More recently, researchers have investigated Type 2 superconductors. A few of these are pure elements such as technetium and niobium, but most are highly complex compounds. Type 2 superconductors have higher T_C , the highest discovered so far being 138 K.

Materials of zero resistance have many practical applications including power lines and high-powered electromagnets. Because there is no resistance, no heat is generated and no energy is wasted. Often the 'wire' is a superconductive tape. This consists of a strip of metallic substrate (such as a nickel alloy) coated with a film of Type 2 superconductor, such as yttrium-barium-copper oxide. This has about 100 times the current carrying capacity of copper wire of equal dimensions. As a cable the conductor has channels for cooling it with liquid nitrogen. Such conductors can carry 600 000 A/cm^2 at 77 K. 1 kg of superconductor cable can replace over 70 kg of conventional copper cable. They are used for building motors, generators, transformers, and electromagnets, all of which operate with very high efficiency because of their no-resistance windings.

Superconducting magnets are used to levitate vehicles above a metal track so that its movement is practically frictionless. In Japan, a MAGLEV train hovers in the air above its track when its superconducting magnets are energized. The MLX01 test vehicle has achieved a world record speed of 1235 km/h while running on the Yamanashi test line. The ability to generate strong, stable magnetic fields has applications in frictionless bearings. Gyroscopes are used as references for orientation of spacecraft but the problem has always been the size and mass of the gyroscope wheel relative to the payload. By using frictionless bearings, a gyroscope with a small, relatively light, wheel turning at very high speed can achieve the required stability.

The strong magnetic fields produced by superconductor electromagnets have other applications. For instance, they have been used to produce the force required for aircraft catapults and for accelerating roller-coasters. They are also used in magnetic resonance imaging (MRI). In a strong magnetic field, the hydrogen atoms in the water and fat of the body accept energy. They release it at detectable frequencies, allowing the distribution of body tissues to be mapped without surgery.

Another application of superconducting magnets is in the storage of energy. Energy may be stored by building up a high-intensity magnetic field. The energy is released later when current is drawn from the magnet and the field collapses. With the very high field strengths that are attainable with superconducting magnets, it is possible to store large quantities of energy in a relatively small volume. Storage is 100% efficient because there are no heat losses and all the stored energy is recoverable.

A more recent use for superconduction has been found in **fault limiters**, devices which, for example, switch off the current at a power station when a fault occurs on the distribution lines. The advantage of superconductor fault limiters is that they act extremely rapidly. They take only a few milliseconds to switch currents of thousands of amperes.

At the other end of the scale, a **superconducting quantum interference device** (SQUID) consists of a superconducting coil which, owing to its zero resistance, is able to produce an appreciable induced current when placed in a very weak magnetic field. SQUIDs have many applications for detecting weak fields or the fields produced by weak currents. For example, a SQUID can detect the electrical field generated by the human heartbeat, and could detect at a distance of 10 km an alternating current amplitude of 1 A flowing in a straight wire. Superconductors are also used in highly sensitive light detectors, for infra-red telecommunications and infra-red astronomy. Sensors based on niobium nitride are able to detect a single photon and operate at 25 GHz.

Internal resistance

Even when a cell is not connected into a circuit, there is an e.m.f. between its terminals. The size of the e.m.f. depends on which two metals are used for the electrodes. The e.m.f. causes a p.d. between the terminals. The size of the p.d. equals the size of the e.m.f. when the cell is unconnected.

In the figure, a cell is causing current (conventional current, see p. 14) to flow round a circuit which consists of a resistor, value R. To complete this circuit, there are ions carrying electric charge through the electrolyte of the cell. Like most conductors, the electrolyte has resistance. Taking this into account, we recognize that the circuit has a second resistance, termed the **internal resistance**, r, of the cell.

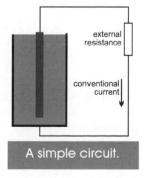

A simple circuit.

The value of *r* depends on the composition of the electrolyte, the distance apart of the electrodes (the central rod and the case of the cell) and their area. It may differ from cell to cell.

In the schematic diagram of the circuit, below, everything on the left is the part of the cell. The circles represent its 'terminals', where the wires of the circuit connect to it.

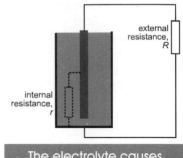

The electrolyte causes internal resistance.

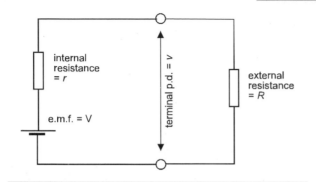

The 'simple circuit' actually comprises *two* resistances, in series with the e.m.f. of the cell.

The e.m.f. of the cell, that is, the actual electrical force produced by the chemical action of the electrodes and electrolyte is *V*. Note that the e.m.f. and the resistance of the electrolyte that produces it are shown *separately* in the diagram.

Suppose that a current *I* is flowing in the circuit in a clockwise direction. The equation on page 25 tells us that the p.d. across the *internal* resistance is *Ir*. So there is a fall of potential as the current flows through the internal resistance. This means that *v*, the p.d. between the terminals, must be less than *V*, the e.m.f. When a cell is connected *into a circuit*, the p.d. between its terminals is always *less than* its e.m.f. This is called the **lost p.d.** across the internal resistance.

The lost p.d. is always equal to *Ir*. If *I* or *r* or both are very small, the lost p.d. is small and the p.d. across the terminals is almost equal to the e.m.f. of the cell. If the circuit draws only a few milliamps, p.d. and e.m.f. are almost equal, but the more current flowing around the circuit, the bigger the difference and the lower the p.d. becomes. This explains why we can not use a battery of zinc-carbon cells to power the self-starter motor of a car. The reason is that zinc-carbon cells have a relatively high internal resistance.

If a zinc-carbon cell is used to supply current to a portable radio set, which takes only a few hundred milliamps, the p.d. remains almost as high as the e.m.f. But, if we try to run the starter motor from such a battery, the large internal resistance of the cells prevents the battery from supplying enough current (several amps) to turn the motor. Most of the e.m.f. of the battery is dropped as a p.d. across its internal resistance, with practically no p.d. across the motor. The battery may get hot because the electrons lose most of their potential energy while passing through the electrolyte, but the motor does not turn. The p.d. across the terminals of the cell is insufficient to power the motor. By contrast, the internal resistance of a lead-acid car battery is very low. It can deliver a large current with relatively little drop in p.d. It also explains why we can obtain an unpleasant electric shock from a 6 V car battery but not from a 6 V torch battery.

4 Capacitance

fter the resistor, the most frequently used electronic component is the capacitor. Its basic structure is very simple. It is like a cheese sandwich, consisting of two metal plates (the bread of the sandwich) with a layer of non-conductor (the cheese) between them. The plates have two wires connected to them so that they can be joined to other components. The material between the plates is called the **dielectric**. Some types of capacitor have air as the dielectric but more often the dielectric is a layer of plastic.

Imagine a capacitor connected across the terminals of a cell (overleaf). Electrons flow from the negative terminal of the cell and enter the plate connected to that terminal. The negative charge on that plate repels the free electrons that are in the metal of the opposite plate. These are the electrons which belong to the outer orbits of the metal atoms but which tend to wander around inside the lattice (p. 8). The electrons flow away, attracted toward the positive terminal of the cell. The atoms they leave behind are now positively charged.

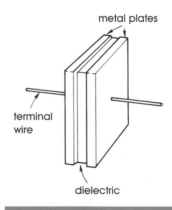

The basic structure of a capacitor.

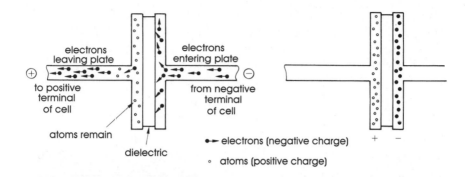

electrons
leaving plate

electrons
entering plate

to positive
terminal
of cell

from negative
terminal
of cell

atoms remain

dielectric

•→ electrons (negative charge)

∘ atoms (positive charge)

+   −

Electrons entering one plate of a capacitor repel electrons from the other plate, which becomes positively charged.

The flow of electrons into one plate and out of the other plate continues until the potential difference (p.d.) between the plates is equal to the p.d. of the cell. Note that, although electrons flow into one plate and out of the other plate, there is no flow of electrons from one plate to the other. Such a flow would be impossible, because the plates are not in contact with each other and the dielectric is a non-conductor.

Suppose that we now disconnect the capacitor from the cell. The plates remain charged as before, and there is still a p.d. between them, equal to the original p.d. of the cell. This illustrates one of the uses of capacitors; they store electric charge (see p. 40). The electrons on one plate remain attracted to the positively charged atoms of the other plate. Once a capacitor has been charged, it holds its charge for a very long time, for hours or even for months. In time, it will become discharged, perhaps by slight leakage of current through the dielectric (it might not be a perfect non-conductor), or when charged ions in the air come into contact with the terminal wires.

Capacitance

The amount of charge a capacitor holds depends on the p.d. between its plates and on its capacitance. Let us consider the p.d. first. If a capacitor holds a certain amount of charge when the p.d. is 1 V, it holds double that amount when the p.d. is 2 V. It holds a hundred times that amount when the p.d. is 100 V. Charge is proportional to the p.d. between the plates.

There is an upper limit to how much the charge can be increased by charging it to a high p.d. This is reached when the p.d. is so strong that the electric field between the plates causes the dielectric to break down. Sparks pass through the dielectric, destroying the capacitor. But most capacitors can be charged to a p.d. of 100 V and some can withstand 1000 V or more.

Capacitance can be understood best by comparing capacitors with tanks of water. Both tanks on the right are filled to the same level but the wider one holds much more water than the other. Similarly two different capacitors may be charged to the same p.d. (equivalent to the water level), but the one with greater capacitance (equivalent to the wider tank) holds more charge (water).

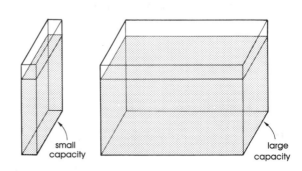

Although they are both filled to the same depth, the two tanks hold very different volumes of water.

We can express these thoughts in a simple equation. Since the amount of charge is proportional to p.d. and to capacitance, the equation is:

$$Q = VC$$

In this equation, Q is the charge, in coulombs (p. 20), V is the p.d., in volts and C is the capacitance, in **farads**. For example, if a capacitor has a capacitance of 2 farads and is charged until the p.d. between its plates is 5 V, then the amount of charge it is holding is $Q = 5 \times 2 = 10$ coulombs. This is rather a lot of charge; charging such a capacitor to that level requires a current of 1 A to flow for 10 seconds.

Practical capacitors

The capacitance of a capacitor depends on four main factors:

- The **area** of the plates: the larger the area, the greater the capacitance.
- The **distance** between the plates: the closer the plates, the greater the capacitance.

Submultiples of the farad

The farad (symbol, F) is too large a unit for practical use. More often we find capacitances quoted in microfarads (symbol μF) where 1 F equals a million microfarads. We also use even smaller units:

1μF = 1000 nanofarads (symbol, nF)
1nF = 1000 picofarads (symbol pF)

- The **overlap** of the plates: the more they overlap, the greater the capacitance.
- The nature of the **dielectric**.

Many capacitors have the structure shown below, in which the plates are made from thin metal foil separated by sheets of plastic film. The plastic sheet is thin, so bringing the plates close together and increasing the capacitance. The plates have a large area but are rolled together to produce a compact device with relatively high capacitance. The whole is encapsulated, usually by dipping it in fluid plastic, which later hardens.

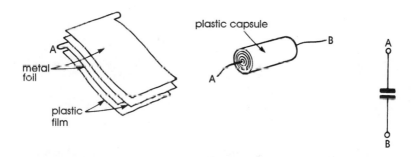

A polystyrene capacitor is made from rolled foil and film. The symbol for a capacitor is on the right.

There are many variations on this theme. In some types, the plates are made by depositing a metal film on a thin plastic foil; the plastic acts as the dielectric.

Capacitors are also made from thin sheets of mica, with silvered surfaces; these have low capacitance but very high stability and are suitable for precision oscillators and tuning circuits. Many different kinds of dielectric are used, conferring special features on the capacitor, such as the ability to withstand high voltage, high capacitance in small volume, high stability in changing temperatures, or suitability for high-frequency operation.

Danger from capacitors

A disconnected charged capacitor holds its charge for a long time. For this reason, a capacitor may retain its charge long after a piece of equipment has been switched off. It is able to deliver the charge very quickly when you touch its terminal wires, producing a dangerously, possibly fatally, high current. Even if a capacitor has apparently been discharged by briefly touching its terminal wires together, it may still retain some of its charge in the dielectric. This residual charge is later transferred to the plates, and could still deliver a powerful electric shock. For this reason, a large-capacitance capacitor should always be stored with its terminal wires twisted together, to ensure that it is fully discharged.

For large values of capacitance, the most frequently used type is the **aluminium electrolytic capacitor**. The plates are made of aluminium and are rolled together for compactness. They are separated by thin sheets of paper soaked in an electrolyte. The main function of the paper is to separate the plates, not to act as a dielectric. In fact, because of the electrolyte, a current can readily pass through the paper. The dielectric is formed by applying a p.d. to the plates. This causes a very thin layer of aluminium oxide to form on the anode plate. This layer is non-conducting and acts as the dielectric. The soaked paper, which is conductive, acts as part of the cathode plate. The oxide dielectric layer is exceedingly thin, so the plates are, in effect, very close together and capacitance is high. Before the anode is oxidized its surface is roughened, which increases its surface area and makes capacitance even higher. This combination of close spacing and large area makes it possible to achieve capacitances of tens of thousands of microfarads in a capacitor of convenient dimensions. Aluminium electrolytic capacitors have wide tolerance, typically ±20%, but sometimes greater. Their capacitance falls if they have not been in use for several weeks, but is gradually restored by subsequent use. The lack of precision and stability makes electrolytic capacitors unsuitable for tuned or timing circuits.

The main disadvantage of aluminium capacitors is that they are polarized, which again limits the kinds of circuit in which they can be used, They must always be connected with the anode positive of the cathode, otherwise the oxide layer is eventually destroyed and the capacitance is lost. The case is marked to show the correct polarity. Leakage current in electrolytic capacitors is appreciably higher than that of capacitors of other types.

Another type of high-capacity capacitor consists of a sintered block containing particles of tantalum. For their size, these **tantalum bead capacitors** have a high capacitance but their breakdown voltages are low and, like the aluminium type, they are damaged by reverse p.d.s. Modern sub-miniature aluminium electrolytic capacitors have almost the same high capacity in small volume as the tantalum beads.

Variable capacitors are often used in tuning circuits in radio sets. The dielectric is air. There are two sets of plates, one fixed and the other mounted on the rotating spindle. The fixed plates are electrically connected to each other and they alternate with the set of movable plates which are also connected to each other. The amount of overlap between the two sets, and therefore the capacitance, is adjusted by turning the spindle. The shape of the plates is such that equal turns of the spindle result in equal changes in capacitance. With all except the largest capacitors of this type, the capacitance is only a few hundred picofarads. Part of the reason for this is that the plates are relatively far apart. This is necessary because the plates may become slightly bent in use, with a risk of touching.

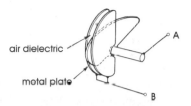

A radio tuning capacitor.

This miniature trimmer capacitor, only 7.5 mm in diameter, has a capacitance adjustable from 1.8 to 22 pF.

A trimmer capacitor has its two sets of plates separated by thin layers of insulating plastic. The area of overlap is adjusted by using a screwdriver to rotate one of the sets of plates. These preset capacitors are adjusted once and for all to set the capacitance in a circuit.

Most types of capacitor are also available as **surface mount** devices. The smaller kinds are made in the 1206, 0805 and 0603 packages (see p. 29).

Charging a capacitor

A capacitor is fed from a constant source of p.d. through a resistor. At every stage in what follows, the p.d. across the resistor is given by:

p.d. across resistor = p.d. of source – p.d. across capacitor

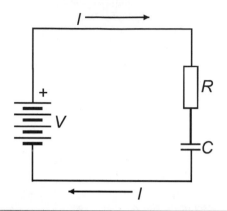

Another equation that holds true at every stage is derived from the equation on p. 25:

In this circuit, the capaicitor is being charged by a *constant voltage* source.

$$\text{current through resistor} = \frac{\text{p.d. across resistor}}{\text{resistance}}$$

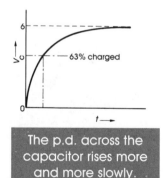

63% charged

The p.d. across the capacitor rises more and more slowly.

The graph shows what happens to v_C, the p.d. across the capacitor as the capacitor becomes charged. The capacitor is uncharged to begin with so the p.d. is zero. Current flows through the resistor and into the capacitor; charge builds up rapidly and the p.d. across the capacitor begins to rise. But, as the p.d. starts to increase, and because the p.d. of the source is *constant*, the p.d. across the resistor is reduced (first equation).

The reduction in the p.d. across the resistor reduces the current through the resistor (second equation). As a result, the flow of current into the capacitor is reduced, charge does not accumulate as quickly as before, and v_C, rises less rapidly. The rate of rise of v_C gradually levels out as time passes. Eventually, there is no further increase. This is when v_C has become equal to V, the p.d. of the source. Now there is zero p.d. across the resistor (first equation). There is no current through the resistor (second equation) and charging ceases.

The curve of the rise of p.d. across the capacitor has a definite shape (we say that it is **exponential**), but we do not need to delve into its mathematics. There is a useful way of specifying a key feature of the curve. This is by quoting the **time constant**, which is defined as the time taken for the p.d. to rise to 63% of the source p.d. If t is this time, in seconds, R is in ohms and C in farads, it can be shown that:

$$t = RC$$

Example: If the resistance is 3.3 kΩ and the capacitance is 10 µF, the time constant is $3300 \times 0.000\,01 = 0.033$ s. It takes 0.03s for the p.d. to rise to 63% of the source p.d. The length of the time constant does *not* depend on the size of the source p.d.

Time constant

The larger the resistance and the larger the capacitance, the longer the time constant.

If a capacitor is allowed to become fully charged (by which we mean to have charged to the same p.d. as the source), and is then discharged by replacing the source by a wire, the p.d. across the resistor is v_C to begin with. As a result of this, v_C falls very rapidly at first. But, as the capacitor becomes partly discharged, the p.d. across the resistor is reduced and discharge is slower. The p.d. falls more and more slowly. In theory it never reaches zero, but in practice it reaches zero in approximately 3 time constants.

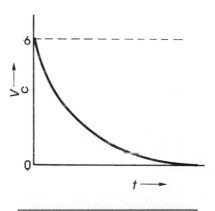

The p.d. across the capacitor falls more and more slowly.

In the example above, the capacitor would take about $3 \times 0.033 = 0.1$ s to discharge to almost zero.

Stray capacitance

Capacitance in a circuit is not limited to that provided by its capacitors. Many other components have capacitance because of their structure. Stray capacitance is also produced when two conducting tracks on a circuit board are very close together, acting as capacitor plates. Such capacitance usually amounts only to a few picofarads and can be ignored. But in high-frequency circuits, such as microwave circuits, the effects of capacitance are much greater and stray capacitance must be allowed for.

Coupling

The dielectric in a capacitor is a non-conductor, so no actual current can pass through a capacitor. But an alternating *signal* is able to pass freely through a capacitor from circuit A (which might be a microphone) to circuit B (which might be an amplifier). This is explained in Chapter 17.

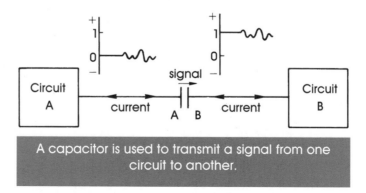

A capacitor is used to transmit a signal from one circuit to another.

The capacitor is coupling the two circuits together. Circuit A on the left is connected to plate A of the capacitor. Assuming that this circuit is a microphone, it generates an alternating potential (though not necessarily a regular sine wave) when it detects sound. When there is no sound, the potential of plate A is zero. When there is sound, the potential alternates about zero, sometimes being positive and sometimes being negative. If the circuit on the right is an amplifier, the average potential at plate B will need to be about 1 V in order for the amplifier to work properly. This potential is provided by biasing resistors (p. 184) not shown in the figure.

During periods of silence, there is an unchanging p.d. of 1 V between A and B. No current is flowing on either side of the capacitor.

When sound is detected, the alternating signal from the microphone causes current to flow into and out of plate A. The potential at plate A oscillates above and below zero, alternately repelling electrons from plate B and attracting electrons into it. In this way, the *signal* (not the electrons) passes *through* the capacitor. Currents flow in and out of plate A and more or less equal currents flow in and out of plate B. The potential of plate B oscillates above and below its constant potential of 1 V. The difference between the potential levels at A and B are shown by the small graphs in the drawing.

In general, capacitors may be used to pass signals from one part of a circuit to another part when the average potentials of the two parts differ.

Decoupling

In the figure below there are two circuits A and B, which take their current from the same power source. The power requirements of circuit A are liable to change very rapidly. Any sudden change in the amount of current drawn from the power line causes the p.d. between the lines to rise and fall suddenly. Such interference with its power supply can upset the action of circuit B. One way to avoid this is to connect a capacitor between the power lines. We use it to isolate circuit B from the effects of circuit A. In other words, it **decouples** A and B.

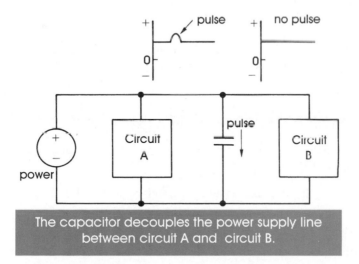

The capacitor decouples the power supply line between circuit A and circuit B.

Decoupling works like this. A sudden change in the p.d. between the lines is the equivalent of a high-frequency signal. Although it may be only one pulse of higher or lower p.d, it is a signal that is easily passed through the capacitor.

The pulse on the power lines in the region of circuit A passes *through the capacitor* from one power line to the other. The capacitor prevents it from reaching circuit B.

Capacitors for batteries

Relatively small capacitors can be made with capacitance as high as 1 F. These are used for storing electric charge as a back-up power supply for the memory in a digital circuit. Developments in the technology of capacitor construction mean that new high-capacity capacitors may well replace chemical cells in many applications. For example, they may be used instead of batteries in electric torches and shavers. One of their advantages compared with rechargeable cells is that they may be recharged very rapidly and no special charger is needed. Another advantage is that capacitors may be recharged thousands of times without any deterioration in their performance. From the point of view of the conservationist, capacitors are preferable to cells because they do not contain toxic metals such as lead or mercury.

New types of capacitor have capacities up to 1.5 kilofarads at 3 volts. A simple calculation shows that the stored charge is $3 \times 1500 = 4500$ C. This means that such a capacitor can deliver an average current of 1 A for 4500 seconds, or one and a quarter hours. An application of interest to automotive engineers is to use the capacitors to deliver very large currents (several hundred amps) for relatively short periods. In this way, they can back up the power supply from the vehicle's conventional lead-acid battery at moments when it is overloaded. In the future, such capacitors may replace the lead-acid battery completely.

Uses of capacitors

Summing up the discussions of this chapter, capacitors are used for:
- Storing electric charge.
- Setting the time constant of timing and oscillating circuits.
- Coupling; transferring alternating signals from one circuit to another.
- Decoupling; preventing disturbances on the power lines from spreading from one circuit to another.
- Backing up the power supply to memory in digital circuits.

5 Inductance

A n inductor consists of a coil of wire, usually surrounding a core of ferromagnetic material. The ferromagnetic material may be iron itself or an iron-containing material known as **ferrite**. An iron core is usually made from layers of sheet iron. Another kind of inductor, is the internal aerial of a portable radio set, which consists of a rod of ferrite on which is wound one or more tuning coils.

Electricity and magnetism

When a current flows through a coil of wire a magnetic field is generated. This is represented by lines of magnetic force, which show the direction in which the needle of a compass points. If there is a ferromagnetic core in the coil, the lines gather to flow through the core. The field is mainly confined to the core, and is stronger. The field is made stronger still if there are many turns and if the current is large.

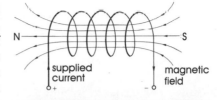

When the current flows as shown, the coil is the equivalent of a bar magnet with its north-seeking pole to the left.

In a current-carrying coil, the *current* generates a *magnetic field*. The reverse effect is found when we take a coil and introduce a magnetic field into it. As the magnet is pushed toward one end of the coil and the field of the magnet begins to enter the coil, a current is generated in the coil. We say that the magnetic field has **induced** a current in the coil.

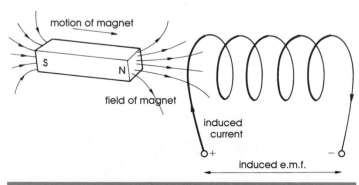

Any change in the magnetic field entering the coil is opposed by the field generated by the induced current (compare with previous diagram).

When the magnet is pulled away, the induced current reverses in direction. The most interesting point about induction is that the current is produced only if the magnetic field is *changing*. A current flows while the magnet approaches and enters the coil. The reverse current flows as the magnet is moved away. There is no flow whenever the magnet stops moving. The *size* of the current induced depends on the *rate of change* of the magnetic field. If we thrust the magnet into the coil rapidly, the current is large. If we move it in slowly, the current is small. Of course, the overall effect is the same, with slow movement a smaller current flows, but for a longer time.

Another key fact about induction is that the direction of the current is such as to oppose the movement of the magnet. This effect is known as Lenz's Law. The induced current in the drawing above flows in the same direction as the supplied current in the previous drawing. The magnet has its north pole toward the coil as it approaches. The field produced by the induced current is in the opposite direction, with the north pole toward the magnet. The two like poles repel each other. We have to do extra work to overcome this repulsion as we push the magnet into the coil. This extra work provides the energy which generates the current. Conversely, as we try to remove the magnet the induced field tries to pull it back into the coil. Once again, we have to do extra work, which appears as the induced e.m.f. that results in an induced current.

Self induction

With the facts of electromagnetism and induction in mind, look again at the diagram on page 49. When there is no current supplied to the coil there is no magnetic field. Now suppose that the current is switched on. Immediately the coil acts as an electromagnet. The effect of this is the same as if we had suddenly placed a magnet inside the coil. A current is induced in the coil to oppose the field already present. This induced current is the result of the field that the coil has itself generated. This effect is known as **self induction**.

The result of self induction is that any change in the amount of current supplied to the coil (turning it on, turning it off, increasing it, or decreasing it) is *opposed* by a self-induced current. If we increase the supplied current, the induced current acts to oppose the increase, to hold the current constant. If we reduce the supplied current, the induced current is in the same direction as the supplied current, acting so as to oppose the decrease, to hold the current constant.

Inductors and inductance

The extent to which an inductor opposes changes in the current passing through it is known as its self inductance. The unit of self inductance is the henry, symbol H. An inductor of 1 H needs many turns of wire wound on a massive core. Most practical inductors have inductances rated in millihenries (symbol, mH) or microhenries (symbol, μH).

The effect of self inductance is proportional to the rate of change of the magnetic field. If the field changes slowly, the induced current is small. If the field changes rapidly, the induced current is high. As a result of this, when a high-frequency signal is applied to an inductor, the inductor acts like a high-value resistance; it blocks the passage of high-frequency signals. Conversely, at low frequencies it acts like a low-value resistance; low-frequency signals pass easily through it. This action is the opposite of the action of a capacitor which passes high-frequency signals but blocks low frequencies (including a fixed d.c. voltage).

We return to this topic in Chapter 16.

Reactance and impedance

Referring back to page 46, we see that capacitors and inductors offer a kind of 'resistance' to the flow of current. Unlike the resistance of a resistor, their 'resistance' depends on frequency. We refer to this as **reactance**, because the effect is produced by the reaction of the capacitor or inductor to the signal being applied to it. Like true resistance, reactance is measured in ohms. But reactance depends on frequency so, when we quote the reactance of a capacitor or inductor, we must state the applicable frequency.

A term that covers both resistance and reactance is **impedance**. The impedance of a component or circuit is the sum of its resistance and both kinds of reactance. It is expressed in ohms, at a stated frequency.

Inductive components

All parts of a circuit have self inductance (or just 'inductance', as we shall refer to it from now on). Even the connecting wires have inductance, though it is small and can be ignored except at very high frequencies. some components have relatively high inductance. These components include:

- Electromagnets: these generate a strong magnetic field and are used for exerting force on ferromagnetic objects. They range in size from the small electromagnet that holds the ejector mechanism down in some kinds of electric toaster until it is time for it to pop up, to large electromagnets used in scrapyards to lift massive objects as car bodies. There are also the electromagnets wound with superconducting coils (p. 34).

- Solenoids: their coil has a ferromagnetic core that can slide easily in and out of the coil (see title photograph, p. 49). Normally, the core is only partly inside the coil. When the current is turned on, the core is drawn forcibly into the coil and pulls on the mechanism coupled to it. Often there is a spring (see upper solenoid) to return the core to its resting position when the current is switched off. They are used for providing force, normally a pulling force, though the narrow rod on the core of the upper solenoid provides a push. Solenoids are used in mechanisms such as electrically operated door locks and fluid control valves.

- Electric motors, relays (p. 128), loudspeakers (p. 125), and electric bells.

- Inductors, described in the next section.

If a circuit contains an inductor or an inductive component such as a relay, it may be subject to unexpectedly high voltages. Suppose that current is flowing through the inductive component, and then it is switched off. The magnetic field in the component collapses immediately. This is a very rapid change and because of this a very large current is induced, such as would be produced by a p.d. of several hundreds of volts. The current flows in the same direction as that in which current was previously flowing, to try to maintain the field. The induced current may be many times greater than that originally flowing, perhaps so large that it damages components such as transistors in nearby parts of the circuit. It also causes intense sparking at switch contacts as they are opened. This may damage the contacts and possibly cause them to become fused together. Because of the effects of inductance, special precautions have to be taken in circuits which switch inductive components rapidly (p. 100).

Practical inductors

Typically, an inductor consists of a coil of wire wound on a ferrite rod. When coated and painted with value markings, it looks just like a resistor. However, it differs from a resistor because it has low resistance (or impedance) to dc (constant) currents and low-frequency signals and it has high impedance to high-frequency signals. Inductors of this type are available in the range 0.1 µH to 100 mH. This type of inductor is used to prevent high-frequency signals from passing. It might be used to decouple (p. 47) a part of a circuit that is oscillating at high frequency to prevent the high frequency from reaching other parts of the circuit, where it is unwanted. An inductor used in this way is called a **choke,** because it chokes or blocks high-frequency signals while allowing low-frequency signals to pass. It might often be connected in a power supply line.

A similar choke effect is often obtained by threading ferrite beads on to the power line conductors. The circular field around the wire creates a circular field in the beads. If this field changes, it induces a current in the conductor to oppose that change.

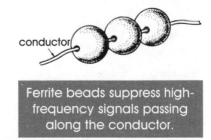

conductor

Ferrite beads suppress high-frequency signals passing along the conductor.

Ferrite beads and cylindrical ferrite sleeves are used to suppress high-frequency interference passing along power lines and along data lines, without affecting the transmission of data.

Another way of suppressing high-frequency signals in a circuit is to wrap a few turns of the conductor around a ring or torus made of ferrite. If turns of two conductors are wound around the same ring they are magnetically linked by the ferrite. Signals from one conductor produce a magnetic field in the ring, which induces signals in the other conductor. This is a simple kind of transformer (see next section).

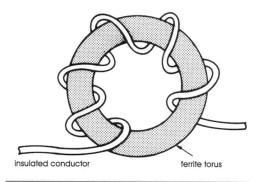

insulated conductor   ferrite torus

High-frequency signals in a conductor may be damped by winding the conductor around a ferrite torus.

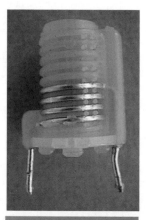

This VHF tuning coil with 3.5 turns of wire has a self inductance of 0.114 μH.

Inductors are used for purposes other than choking. When connected to a capacitor, they form a tuned network that resonates at a particular frequency (p. 66). The inductor may be a few turns or wire wound on a plastic **former,** as on the left.

Tuned inductors may have more than one winding (opposite), and these are equivalent to transformers that operate best at one particular frequency. One or more coils are wound on a plastic former. The iron-dust core is threaded, as is the inside of the former. The core is moved into or out of the coil by turning it with a special non-magnetic tool so as not to introduce spurious magnetic effects. This allows the self inductance of the coil to be adjusted precisely.

If there are two or more coils, the core links them electromagnetically, so they function as a transformer (see next section). Movement of the core in or out of the former adjusts the linking between the coils when setting up the circuit. The inductor is enclosed in an earthed metallic can to prevent magnetic interference between the coils and other adjacent inductors and nearby parts of the circuit.

A very large range of ready-wound coils is available, including versions for use on surface mount circuit boards.

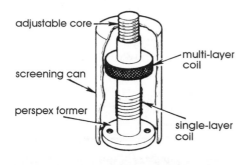

Tuning coils of this type are used in radio-frequency circuits.

A short-wave oscillator coil, 1.1 μH.

Transformers

A transformer is made by winding two coils on a single core. Most often the core is shaped to form a ring, so that all the magnetic field produced by one coil passes through the other coil. The coil to which current is supplied is known as the **primary coil**. When an alternating current passes through this it generates an alternating field in the core. The alternating field induces a current in the **secondary coil**. Note the word 'alternating' in the previous sentence. Current is induced only when a magnetic field is changing (p. 50). A transformer requires alternating current. It does not work with direct current.

The drawing on the right illustrates the *principle* of a transformer, but a practical power transformer has its coils wound on a core made from laminations that are shaped liked squared 'eights'. Or they may be wound on a toroidal core.

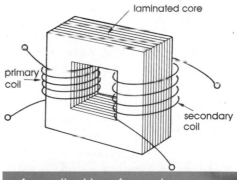

The form of the core is such that the lines of force are confined *within* the core and are not able to spread outward to interfere with adjacent parts of the circuit.

A practical transformer has many more turns in its coils than shown in this drawing. Also, the coils are not wound separately, but one on top of the other.

Eddy currents

The magnetic fields in an inductor may generate electric currents in the core. Such eddy currents represent a waste of energy, leading to the generation of heat and interfering with the proper operation of the inductor. Laminating the core from sheets of iron cut to the same shape is one way of reducing eddy currents.

The action of a transformer depends on the numbers of turns in the primary and secondary coils. The ratio of the number of secondary turns to the number of primary turns is the **turns ratio**, n, where:

$$n = \frac{\text{secondary turns}}{\text{primary turns}}$$

In a **step-up transformer**, n is greater than 1. When an alternating current of given amplitude passes through the primary coil, the alternating current (at the same frequency) which is induced in the secondary coil has an amplitude n times as great. For example, suppose the primary has 10 turns and the secondary coil has 40 turns. The turns ratio is given by $n = 40/10 = 4$. If the current fed to the primary has an amplitude of 2.5 V, the current in the secondary coil has an amplitude of $4 \times 2.5 = 10$ V. Step-up transformers are used at power stations to step up the voltage from the generator to many kilovolts before the power is fed to the distribution network.

If the turns ratio is 1 (primary and secondary turns are equal in number), there is no change in voltage. It might seem that there is little point in having a transformer of this kind, but such transformers are used to isolate one circuit from another. For example, we may have equipment which operates at mains voltage (230 V) but, for reasons of safety, we do not want to have it connected directly to the mains. An isolating transformer with its primary connected to the mains and its secondary connected to the equipment provides the required isolation.

A **step-down transformer** works in the opposite sense to a step-up transformer and has a turns ratio of less than 1. Such transformers are often in use in mains-powered electronic equipment. Devices such as radio receivers, audio amplifiers, electronic keyboards, and personal computers operate on low voltages such as 12 V, 9 V or less. The first stage in providing a power supply for these is to step-down the voltage from 240 V to the level required.

Transformers may also be used for coupling microphones and loudspeakers to amplifier circuits and for coupling different sections of amplifier, radio and oscillator circuits.

Power

The primary coil of a transformer converts electrical energy into magnetic energy. Its rate of working or power is given by this equation:

$$P = IV$$

This power is recovered in the secondary coil, where magnetic energy is converted back into electrical energy. There can be no gain of power during these processes. The principle of the conservation of energy prevents this. In fact there will be a certain amount of power loss, due to heat produced in the coils and heat produced in the core due to eddy currents (opposite). Good transformer design minimizes power loss to a small percentage. Assuming that there is *no* loss (that the transformer is 100% efficient) the value of *P* must be equal on both sides of the transformer. So, if *V* is increased four-fold, for example, then *I* must be reduced to a quarter. The maximum current that can be drawn from the secondary coil is limited to one quarter of that supplied to the primary coil. Usually, the primary coil of a step-up transformer is wound in heavy-duty wire to allow a large current to be supplied, while the secondary coil consists of many turns of fine wire. The reverse applies to a step-down transformer.

6 Simple circuits

Resistors, capacitors and inductors are described as **passive** components. The currents through them and the p.d.s across them are the result of external currents and p.d.s. They contrast with the **active** components such as transistors. Before we go further, we look at ways in which the passive components are connected to make useful circuit building blocks.

Potential divider

This consists of two or more resistors wired in series. In Figure A opposite, we have a potential divider made up of ten equal resistors, resistance 1 Ω each. For each resistor, $V = IR$ (p. 25) and, because the same current flows through all of them and they all have the same resistance, I and R are the same for all, making V the same for all. As we go down the chain, the potential drops by the same amount across each resistor. From top to bottom, it drops from V to zero in ten equal steps, each of $V/10$.

In Figure B there are just two resistors, 3 Ω and 7 Ω. Think of the 3 Ω resistor as being equivalent to three 1 Ω resistors in series. It is equivalent to three of the steps of Figure A. The p.d. across it equals three steps of $V/10$, making a drop of $3V/10$. Similarly, the drop across the 7 Ω resistor is $7V/10$. If $V = 20$ volts, the drops are 6 V and 14 V respectively.

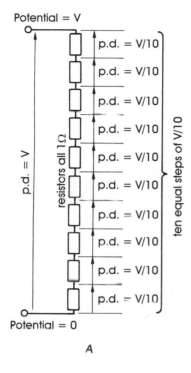

A

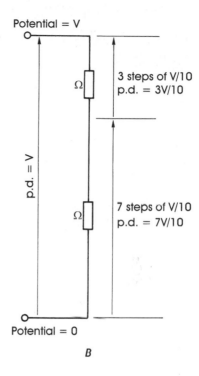

B

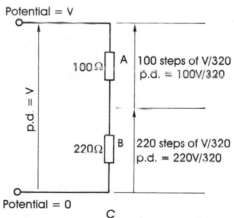

C

A potential divider consists of a chain of two or more resistances. The potential across the chain is divided in proportion to the individual resistances.

The p.d. across either one of the resistors is proportional to its resistance. In Figure C we have two resistors totalling 320 Ω. These are the equivalent of 320 individual resistors, each of 1 Ω. There are now 320 equal steps, and the p.d. per step is $V/320$. The p.d.s across the two resistors are:

$$V_1 = 100V/320 \text{ and } V_2 = 220V/320$$

If the value of V is 16 volts, the p.d.s across the resistors are $(100 \times 16)/320 = 5$ volts, and $(220 \times 16)/320 = 11$ volts. Once again, the p.d.s are proportional to the resistances.

A circuit such as this, which divides the total p.d. into two or more parts, is called a **potential divider**. This can be useful in circuits where, for example, the battery provides a p.d. of 6 V but we need a p.d. of 5 V or some other value less than 6 V. A general rule for finding the potential V_{junc} at the junction between two resistors A and B is:

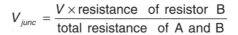

$$V_{junc} = \frac{V \times \text{resistance of resistor B}}{\text{total resistance of A and B}}$$

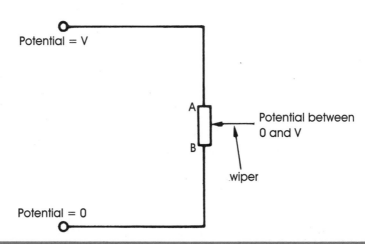

A variable potential divider built from a variable resistor.

A potential divider can be built from a variable resistor. The two sections of the track on either side of the wiper represent two resistors in series. As the wiper is moved along the track the total resistance (and therefore the number of steps and the p.d. per step) remains unchanged. The potential *at the wiper* falls smoothly from V when the wiper is at A, to zero when the wiper is at B. This is a way of obtaining any potential within a given range.

The calculations about potential dividers assume that the same current flows along all the resistors. If another circuit is connected to the divider, this may not be true. If the circuit has very high resistance (left, below) it takes so little current that almost all of the current flowing through A goes on through B. The divider gives the calculated potential. If the connected circuit has low resistance (right, below), most of the current that flows through A goes on to flow *through the circuit*, and only a small current is left to flow through B. An unexpectedly small current through B means that the p.d. across it is much lower than calculated. This effect may be important when making measurements with test-meters.

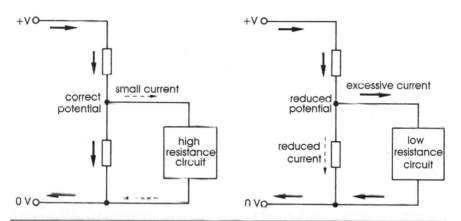

A potential divider delivers the calculated potential only if it is feeding current to a high resistance circuit.

Resistor bridge

A bridge is made by connecting four resistors as in the figure overleaf. It can be shown by thinking of R_A and R_B as a potential divider and R_C and R_D as another potential divider, that if:

$$\frac{R_A}{R_B} = \frac{R_C}{R_D}$$

then the potential at X is exactly equal to that at Y. We say the bridge is **balanced**. The milliammeter shows no current flowing between X and Y. The point about using a bridge is that it is very sensitive to being put out of balance. If any one of the values changes even by a small amount, the bridge loses balance and a current flows through the meter.

This type of circuit is used for measuring resistance, with the unknown resistance at R_A, a high-precision variable resistor at R_B and two high-precision fixed resistors at R_C and R_D. R_B is adjusted until the bridge balances (zero current through the meter). Then, knowing R_B, R_C, and R_D, the value of R_A is calculated. A bridge is often used with resistive sensors, such as strain gauges, and the idea is adaptable to measuring capacitance and inductance.

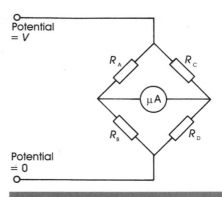

A bridge provides a precise measurement of resistance.

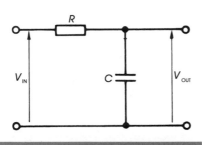

Because the reactance of the capacitor varies with frequency, this circuit has frequency-dependent action.

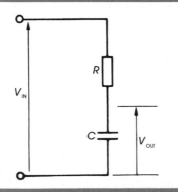

Now it looks like a potential divider.

Filters

The resistor-capacitor network shown on the left is often used in audio and radio circuits and occasionally in others. If its input V_{IN} is constant, C soon charges to a constant p.d. and the output V_{OUT} settles at a value equal to V_{IN}. The circuit does something more useful when the input is an *alternating* p.d., amplitude V_{IN}. To see what happens, we re-draw the diagram. Now it has the same arrangement of parts as a potential divider, but with a capacitor replacing the lower resistor. The reactance (equivalent to resistance, p. 52) of the capacitor depends on the frequency of the signal, the alternating p.d. It is high for low frequencies and low for high frequencies.

If the signal has low frequency, the circuit behaves as if it is a potential divider with a very high-value resistor in place of the capacitor.

Example: Suppose that the impedance (= resistance) of R is 1 kΩ. Suppose also that the capacity of C is 100 nF (nanofarad). If V_{IN} is an alternating signal of 1 kHz with an amplitude of 1 V, the graphs of V_{IN} and V_{OUT} look like this:

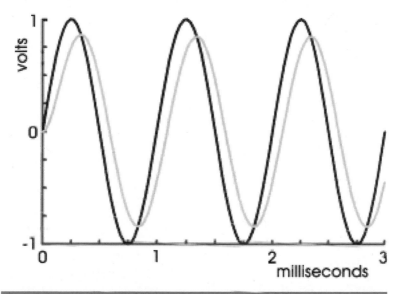

A signal alternating at 1 kHz passes through the resistor-capacitor network, with little reduction in amplitude.

The signal V_{IN} (black curve) with a frequency of 1 kHz takes 1 ms to complete a cycle. The output signal V_{OUT} (grey curve) has *exactly* the same frequency, and its amplitude about 840 mV. It has passed through the network with only a small amount of loss. This is because C has a relatively high impedance *at that frequency*. It is behaving as if it is a resistor of relatively high value.

Frequency

The unit of frequency is the hertz, symbol Hz. A signal alternating at 1 Hz takes 1 second to go though a complete cycle. Units of higher frequency are kilohertz (kHz, thousands of cycles per second), megahertz (MHz, millions of cycles per second) and gigahertz (GHz, thousands of millions of cycles per second).

This is what happens if we keep the same components values but increase the frequency of the signal by ten times to 10 kHz:

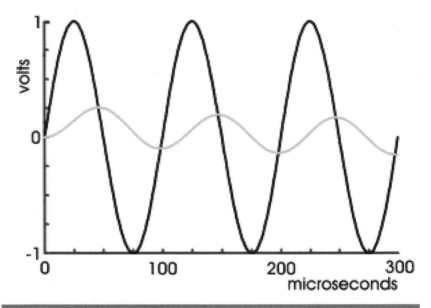

A signal of 10 kHz is much reduced in amplitude when it is passed through the network. Note the time scale of this graph is only a tenth of that of the graph on the previous page.

The grey curve shows that the amplitude of V_{OUT} is now reduced (or **attentuated**) to about 157 mV. This is explained by the fact that the impedance of the capacitor is reduced as frequency is increased. Now only a relatively small p.d. can develop across it. Note that the frequency of V_{OUT} is still equal to that of V_{IN}.

The graphs illustrate the action of the network at frequencies of 1 kHz and 10 kHz. Below 1 kHz the impedance of C is even higher than at 1 kHz and the amplitude of V_{OUT} is very close to 1 V. At an extremely high frequency the impedance of C is very small. It is almost a short-circuit and V_{OUT} is very close to zero. The overall effect is that low-frequency signals appear strongly at the output of the network, but high-frequency signals are lost. We describe this network as a **low-pass filter**. More exactly, frequencies below a certain level, known as the **cut-off frequency**, pass through the filter with very little attenuation. But there is progressive reduction in amplitude above the cut-off frequency. The way the amplitude varies with frequency is shown in detail on page 202.

Because it uses only passive components (see the start of this chapter), this resistor-capacitor network is more aptly described as a **passive low-pass filter**. Later, we shall describe some active filters.

Filtering affects not only the amplitude of the signal at different frequencies but also the timing or **phase** of the output signal with respect to the input signal. This is because of the time factor involved in charging and discharging the capacitor. The sine wave of the output signal is still a sine wave and has the same frequency as the input signal, but its cycles begin and end a little bit later than those of the input signal. The delay in the graph on page 63 is approximately 90 μs, or just under one tenth of a cycle. At the cut-off frequency (1.6 kHz for this network), the output signal is an eighth of a cycle behind the input signal. As frequencies increase further, the output signal lags further and further behind the input signal until it is a quarter-cycle behind.

Another type of filter is obtained if we exchange the positions of the resistor and the capacitor in the network. Now high-frequency signals pass freely through the circuit but low-frequency signals are much attenuated. This is a **high-pass passive filter**.

This simple description is only an outline of the working of resistor-capacitor filters. We have seen that a filter affects not only the amplitude of the output signal but also its phase in relation to that of the input signal. The same principles apply to a filter made from a resistor and inductor or from a capacitor and an inductor. As might be expected, a filter like that on page 62, but with an inductor instead of the capacitor has the opposite response to frequency. It is a high-pass filter.

Resonance

This circuit is very sensitive to the effects of frequency. If the input to the circuit is a low-frequency signal, it is mainly blocked by the capacitor. But it passes easily through the inductor, which acts as a short circuit. The impedance of the capacitor-inductor network is very low impedance and only small p.d.s are developed across it.

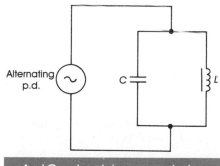

An LC network is resonant at a particular frequency.

If the input to the network is a high-frequency signal, it passes readily through the capacitor but not through the inductor. As before, the impedance of the capacitor-inductor network is very low impedance and only small p.d.s are developed across it.

However, at one *intermediate* frequency, depending on the values of capacitance and inductance, the capacitor and inductor offer equal reactance to the signal. Then the signal has maximum effect on the circuit, and the swings of p.d. across it are at a maximum. When the signal source is oscillating at this frequency, the circuit is said to **resonate**.

At the resonant frequency it is not necessary to supply large quantities of energy to make it oscillate. In fact, even if the signal source is removed, it will continue to resonate for a while, until the oscillations die out. The action is very like that of pushing a child on a swing. The swing has a natural frequency which depends on its length. As the child swings to and fro we give it a small push every time the child is swinging away from us. Gradually, the amplitude of swinging increases, because each time we push we are reinforcing the natural motion of the swing. When the person is at the extremes of the swinging, at the highest points above ground, energy is stored in the person as potential energy. When the person is at the lowest point of the swinging the potential energy is gone, and is converted to energy of motion. As the swing oscillates to and fro, energy is converted from potential energy to energy of motion (kinetic energy) and back again repeatedly.

A capacitor-inductor circuit is like a swing. Imagine the capacitor fully charged in a given direction. It is storing energy in the form of the electric field produced by the stored charge. Then the capacitor discharges through the inductor, losing energy itself but building up a store of energy in the inductor in the form of its magnetic field. At a certain point the capacitor is fully discharged and the inductor holds its maximum energy. With no further current from the discharged capacitor the magnetic field in the inductor begins to collapse. However, self induction (p. 51) ensures that a current continues to flow through the inductor in the same direction as before. This recharges the capacitor but in the opposite direction. Eventually the field has completely collapsed and a charge has built up in the capacitor. Once again all the energy is stored in the capacitor.

At this point, the capacitor is fully charged but in the opposite direction. With no further change in the magnetic field, no more current flows into the capacitor. It begins to discharge through the inductor and the process repeats, but with charge, currents and fields in the opposite direction. There is repeated conversion of energy from electric field to magnetic field and back again.

If there were no loss due to resistance of the wires of the inductor and connections, and if there were no small losses in the dielectric of the capacitor, the oscillations described above would go on for ever.

A resonant circuit can build up large swings of p.d. from very small beginnings. Starting with a minute amount of energy in the electric or magnetic field, small oscillations will occur as described above. If an external source supplies energy from outside at just the right frequency (like the person pushing the swing at just the right instant), a little extra energy is added to the system at each oscillation. The energy levels build up gradually until the circuit is oscillating strongly. This is the principle on which certain types of oscillators work (p. 177) and also that of the tuning circuits of radio receivers (p. 273).

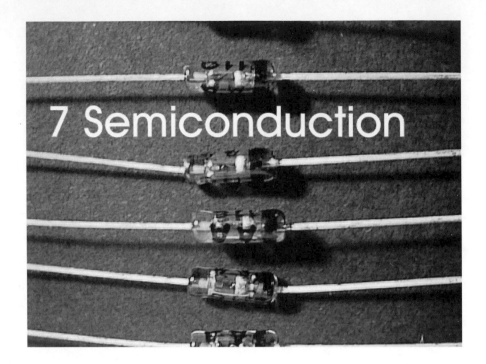

7 Semiconduction

In Chapter 1 we explained that an electric current consists of a flow of charge carriers. In most cases, the charge carriers are electrons, which are negatively charged, but it is also possible for charge to be carried by ions, which may be charged either negatively or positively. The requirements for the conduction of electricity are:

- There must be a supply of charge carriers, such as electrons, or ionized atoms or molecules.

- The charge carriers must be free to move; electrons in the lattice of a metallic crystal, ions in solutions or in near-vacuum conditions.

- There must be an electric field to provide the force (e.m.f.) to move them.

Apart from the superconductors (p. 33) below their critical temperatures, the best conductors are the metals. This is because the atomic structure of a metal is such that its lattice has a good supply of freely-mobile electrons wandering within it (p. 8). Carbon is a non-metal, but this too is a good conductor. At the other extreme are the non-conductors, including most plastics and ceramics, which have no free electrons.

Between the conductors and non-conductors lies a class of materials known as the **semiconductors**.

Semiconductors were given that name because they are non-conductors at low temperatures but are conductors at room temperature and above. Two of the most widely-used semiconductors are the elements silicon and germanium. Most of the descriptions which follow refer to silicon but, except where differences are pointed out, it can be taken that the description applies equally to germanium. The drawing below is a view of the lattice of a crystal of silicon. It is shown in only one plane, instead of three dimensions, to make the drawing simpler.

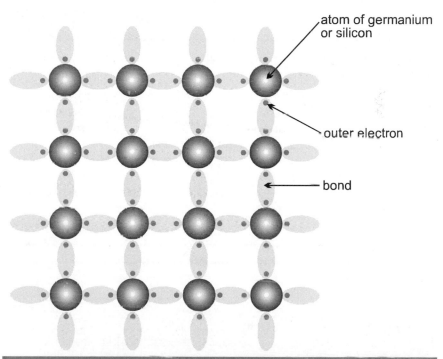

The structure of the lattice of a pure semiconductor shows the atoms bonded by sharing electrons with four adjacent atoms.

The electrons of an atom occur in **shells** (concentric layers) surrounding the nucleus. Each atom of silicon has four electrons in its outer shell. The ideal number of electrons to fill this shell is eight. Each atom shares an electron with its four neighbours. Though it has only four electrons at any given instant, the shared electrons circle in the outer orbits of each atom and its neighbours, randomly going from one atom to another.

At different times each atom has a full outer shell. This creates a bond between the adjacent atoms, holding the atoms together in the lattice. At room temperatures and above, a few of these electrons acquire additional (thermal) energy. Their velocity increases and they escape from their orbits. They are then free to wander in the lattice (p. 8) and to act as charge carriers. The semiconductor conducts but, because there are relatively few free electrons in semiconductors, they do not conduct as readily as metals.

The electrons released from the atoms of the semiconductor are known as **intrinsic charge carriers** and conduction by these is called **intrinsic conduction.**

Resistance and temperature

Increasing temperature causes more electrons to escape from the atoms, so that the conductivity of a semiconductor increases with temperature. Its resistance *decreases*. By contrast, a conductor such as a metal already has its complement of free electrons at any temperature. Heating the metal does not release more electrons. Instead it increases the extent of the vibrations of the atoms. The vibrating atoms impede the flow of electrons through the lattice. Its resistance *increases*.

Thermal runaway

A piece of semiconductor that is carrying a current becomes warmer, so decreasing its resistance. This allows the current through the semiconductor to increase, which results in further warming, further decrease in resistance, and further increase of current. This process may accelerate until the semiconductor becomes so hot that it melts. This is known as thermal runaway. It is avoided by designing circuits so as to limit currents to safe levels or by providing a heat sink. This is a plate of heavy-gauge copper or aluminium, often with fins, that distributes the excess heat. The heat sink is bolted to or clipped on to the device, their contacting surfaces being coated with a paste that helps conduction of heat from the semiconductor to the heat sink.

Increased conduction

Conduction by a semiconductor is increased by doping it (see box). The effect of this is to introduce a very small percentage of dopant atoms into the lattice. Antimony is one of the elements used for doping silicon. The antimony atom has five electrons in its outer shell, one more than is found in silicon. This provides a spare electron, which is free to become a charge carrier and so to increase the conductivity of the silicon.

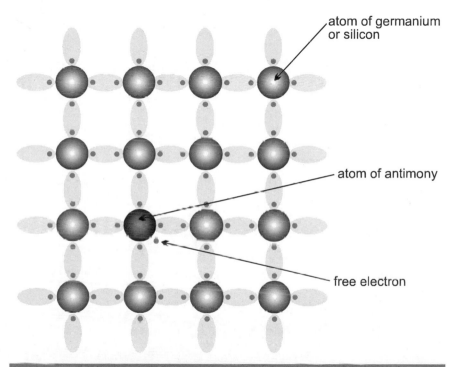

atom of germanium or silicon

atom of antimony

free electron

Doping silicon or germanium with antimony provides a free electron, which increases conductivity.

Another element used for doping silicon is phosphorus; this too has five electrons in its outer orbit. Charge carriers introduced by doping are called **extrinsic charge carriers**, and conduction by these carriers is called **extrinsic conduction**. Silicon that has been doped with carriers such as antimony or phosphorus, which provide negative charge carriers (electrons), is called **n-type silicon.**

Doping

A doped semiconductor consists of the pure substance in crystalline form into which certain impurities (dopants) are diffused in extremely small quantities. Roughly cylindrical crystals of silicon are grown up to 30 cm diameter and 1.5 m long, then sliced into circular wafers. A whole wafer of silicon is doped by heating it in an oven and passing over it a vapour containing the dopant. The dopant condenses on to the surface of the silicon. The atoms of dopant diffuse into the silicon, eventually becoming evenly spread throughout the wafer.

Another way of providing more charge carriers is to dope the silicon with an element such as indium or boron, which have only three electrons in their outer orbits. This results in a 'missing' electron at the site of each atom of dopant. We refer to this 'missing' electron, or 'vacancy' as a **hole**.

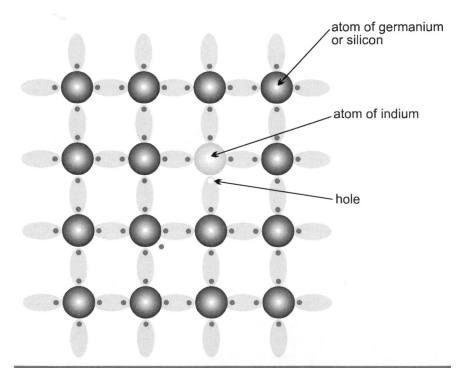

Doping with indium provides a hole, which increases conductivity.

The hole may soon be filled with an electron that has escaped from a nearby silicon atom. The escape of that electron creates a hole, which may later be filled by an electron escaping from elsewhere in the lattice. The overall result of doping is that there is a shortage of free electrons so there are more holes than free electrons.

What happens when an electric field is present is shown below:

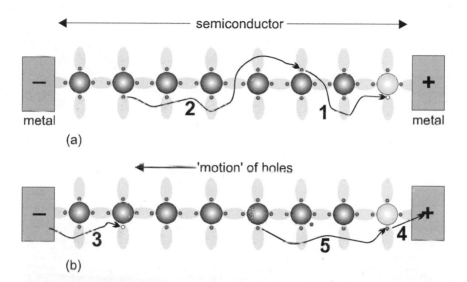

Conduction in p-type silicon. Holes move from positive to negative and so act as positive charge carriers.

The first stage is illustrated in (a). At (1) an electron escapes from a silicon atom and drifts in the electric field until it finds a hole at the indium atom. At (2), the hole this created is filled by an electron escaping from an atom further along the material. The electrons drift to the right (toward +), while the vacancies or holes pass from right to left (toward −). The next stage is illustrated in (b). At 3, the hole created at (2) is filled by an electron from the negative supply, having travelled along a metal wire. At 4 an electron escapes and passes into the metal wire of the positive supply. The hole created is again filled at 5 by an electron, so shifting the hole toward the negative end.

There is no 'flow' of electrons in the semiconductor. They merely jump from one atom to another atom a very short distance away. Because of the electric field, they fill a hole that is in the direction of the positive end. The effect of this is that the holes move in the opposite direction, toward the negative end.

Whenever an electron fills a hole this results in a hole being created further along, in the direction of the negative supply. In effect, the holes move toward the negative end of the semiconductor. This is the direction in which positive charge carriers move, so we can consider holes to be positive charge carriers. For this reason, silicon doped with indium or boron is known as **p-type silicon**.

The p-n junction

One of the most important effects in electronics occurs when we have n-type silicon in contact with p-type silicon. This is called a **p-n junction**. Such a junction can be made by taking a bar of n-type silicon and placing a disc of indium against half of it. Heating in a furnace causes some of the indium to diffuse into the silicon. Another method is to take a slice of n-type silicon, and heat it in a furnace with one surface exposed to boron vapour. The amount of indium or boron diffused into the n-type silicon produces enough holes to take up all the free electrons in that region of the silicon, with holes to spare. The result is that the indium-doped or boron-doped region becomes p-type silicon.

Consider a bar of silicon, doped so that half is p-type and half is n-type. The bar is not connected to a circuit, so no external electric field is applied to it. Electrons are free to wander at random.

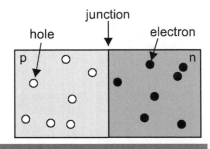

Initial p-n junction.

Some of the electrons from the n-type material wander across the junction, attracted by the holes in the p-type material. The holes in the region of the junction become filled. In the p-type region (to the left of the junction) the filling of holes means that the atoms have, on average, more electrons than they should. The atoms have become negative *ions*.

Similarly, in the n-type region the atoms have, on average, lost electrons leaving holes and so have become positive *ions*.

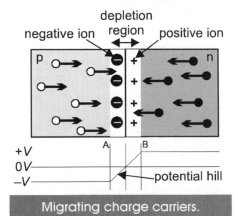

Migrating charge carriers.

Although these ions are charged, they can not act as charge carriers. This is because they are *fixed* in their places in the lattice. The result is that the region on side A of the junction has overall negative charge and the region on side B has overall positive charge. There is a p.d. across the junction. The effect of this p.d. is the same as if a cell was connected across the junction. This is not a real cell but we refer to its *effects* as a **virtual cell**. At a silicon p-n junction, the p.d. is about 0.6 V. At a germanium junction, it is about 0.2 V.

As more and more electrons wander across the junction and fill the holes on the other side, the charged regions A and B become wider. Eventually, the negative charge in region A prevents more electrons from being attracted through to the p-type material, and the charged region does not become any wider. It contains no charge carriers and, for this reason, it is called the **depletion region**. We think of the p.d. across the depletion region as a **potential hill**, the potential rising as we pass from the p-type side to the n-type side. The hill is so steep that electrons can no longer pass down it.

The situation described in the paragraph above is changed if the semiconductor is connected into a circuit and a p.d. is applied to it from an external source. We call this **biasing** the junction. We can bias the junction in either direction.

With **reverse bias**, the external source (such as a dry cell) is connected so that its p.d. is in the same direction as the p.d. of the virtual cell. It reinforces the action of the virtual cell, making the potential hill higher. This makes the depletion region wider than before, and there is even less chance of carriers finding their way across it. No current flows across the junction.

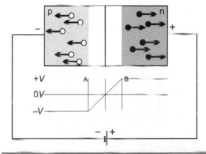

Reverse bias — current blocked.

If the external source is connected so that its p.d. opposes that of the virtual cell, the junction is **forward biased.** The depletion region becomes narrower. The potential hill becomes lower. Electrons are more easily able to cross the junction.

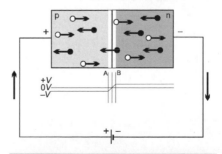

Forward bias — current flows.

If the external p.d. is greater than 0.2 V (for germanium) or 0.6 V (for silicon), the depletion region disappears and charge carriers flow freely through the material.

From this account, it is seen that the p-n junction has the unusual property that it allows conduction in one direction but not in the other.

Semiconductor diode

A semiconductor device that has a single p-n junction is known as a **diode**. The behaviour of a diode is investigated by connecting it across a variable source of p.d. and measuring the current which passes through it.

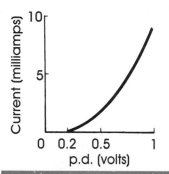

Current through a forward biased diode does not obey Ohm's Law.

The graph shows what happens when the diode is forward biased. As the p.d. increases from zero to 0.2 V (for germanium) or to 0.6 V (for silicon) the p.d. is insufficient to overcome the virtual cell at the junction. No current flows. As the p.d. is further increased, the virtual cell is overcome and current begins to flow. Current increases with increasing p.d., but the graph is not a straight line. This is because flow through a diode does not obey Ohm's Law (p. 25). Note that current is plotted on a milliamp scale.

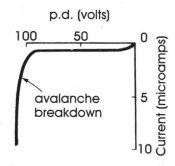

Current through a reverse biased diode is very small until avalanche breakdown.

If the diode is reverse biased, the external p.d. reinforces the p.d. of the virtual cell and the depletion region becomes wider. Only an exceedingly small reverse current flows. This is plotted in *micro*amps on the graph. This leakage is carried by minority carriers, electrons or holes resulting from impurities in the semiconductors. Most diodes withstand high reverse p.d.s, of 100 V or more without any increase of leakage current.

At higher reverse p.d.s, the minority carriers are strongly accelerated by the field. They gain energy and, passing close to atoms in the lattice, cause them to lose electrons. This creates a new supply of free electrons and holes. These too are accelerated strongly, interacting with the atoms to free more electrons and create more holes. The effect is like an avalanche. The current builds up very sharply. Unless the diode is designed to withstand such a large current, it is destroyed. The maximum p.d. that a diode can withstand when reverse-biased is known as the **peak inverse voltage** (PIV).

Current through a diode

The flow of current through a forward-biased diode may be summarized like this. Electrons pass from the external circuit and enter the n-type material of the diode. They flow through the n-type material as far as the p-n junction. Electrons pass from the p-type material into the external circuit, creating holes. These holes flow through the p-type material as far as the p-n junction. The holes are then filled with electrons from the n-type material. Thus, although electrons enter and leave the diode, charge is carried by holes in the p-type region.

Diode valves

The name *diode* arose in the days before semiconduction was discovered, when almost all circuits were constructed using valves (vacuum tubes). One type of valve, called a *di*ode because it contains *two* electrodes, has the property of one-way conduction. The glass envelope or tube is sealed and contains a vacuum. It also contains a **cathode**, usually in the form of a thin filament of wire, and a metal plate called the **anode**.

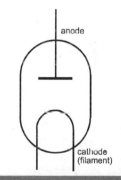

anode

cathode
(filament)

The symbol illustrates the essential features of a diode valve.

The filament is heated to red heat by passing a current from a low-voltage supply through it. This causes the metal atoms to emit electrons, which form a cloud around the cathode. When a p.d. is applied between the cathode (negative) and anode (positive), electrons are attracted toward the anode.

They carry a current through the diode from cathode to anode. Electrons are set free only at the hot cathode, so the diode does not conduct when a reverse p.d. is applied.

The name of the diode has passed to its semiconductor equivalent. Valves are less often used today except in specialist applications such as high-power radio transmitters and certain designs of hi-fi audio amplifier.

Practical diodes

The title photograph of this chapter shows typical low-power diodes. They consist of a glass or plastic capsule usually a few millimetres long containing a bar of silicon (or germanium in the case of the diodes in the photograph), in which there is a p-n junction. Wires are connected to the n-type and the p-type materials. These connections are called **cathode** and **anode** respectively, the names being the same as the two electrodes of the diode valve. As in the valve, electrons flow from cathode (n-type) to anode (p-type). Conventional current (p. 14) flows from anode to cathode. Often, the cathode end is indicated by a band marked on the capsule. In the photograph, the diodes have their cathodes to the right, so current flows from left to right.

Diodes are used for many purposes, wherever current must be allowed to flow in only one direction (p. 76). Typically, the current carried by such diodes is rated in milliamps or less. One special type of low-power diode is the **signal diode**, used in radio and TV detection circuits (p. 274) and able to respond to alternating p.d.s changing at radio frequencies. Signal diodes may also be used in high-speed logic circuits.

A **rectifying diode** is intended to carry currents rated in amps. It has a cylindrical plastic case, usually rather larger than that of a signal diode. Power diodes have a metal case with a threaded shaft at one end so that it can be bolted to a heat sink (p. 70). Rectifying diodes take their name from the fact that they are used in the rectifier sections of power-supply circuits (p. 104).

Light-emitting diodes and laser diodes are described in Chapter 13.

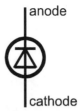

anode

cathode

The symbol for a diode indicates the direction in which current flows.

Zener diodes

Many diodes sold as 'Zener diodes' are actually **avalanche diodes**. They rely on the avalanche effect described above, but are specially designed so that the avalanche action begins at a sharply defined reverse voltage, and so that the resulting high current does not destroy the diode. If we apply an increasing reverse p.d. to such a diode, there is no current (except for a minute leakage current) until the avalanche p.d. is reached, when a large current begins to flow. These diodes have applications in voltage regulator circuits (pp. 105, 108). Avalanche diodes are made so that the effect occurs at a specified reverse p.d., usually in the range 5 V to 200 V.

The true Zener diodes, which have the same properties as avalanche diodes and are used in the same applications, rely on the **Zener effect**. The diode is made from heavily doped silicon, so that the depletion layer is very thin. Its potential hill is very steep. When the reverse p.d. exceeds a certain value (in the range 2.5 V to 5 V), electrons are able to tunnel through the depletion layer and current begins to flow.

Variable capacitance diodes

A reverse-biased diode has two conducting regions (n-type and p-type) with the depletion region between. This may be compared with the structure of a capacitor. Varying the reverse p.d. alters the width of the depletion region, and this alters the capacitance. In other words, this is a voltage-controlled capacitor.

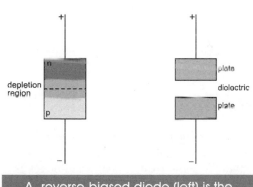

A reverse-biased diode (left) is the equivalent of a capacitor (right).

Such diodes are usually sold under the name of 'varicap' diodes. Their capacitance is small, usually no more than a few hundred picofarads, and they are used in certain kinds of radio and TV tuning circuits. They have the advantage that the tuning can be controlled electronically instead of by turning a tuning knob.

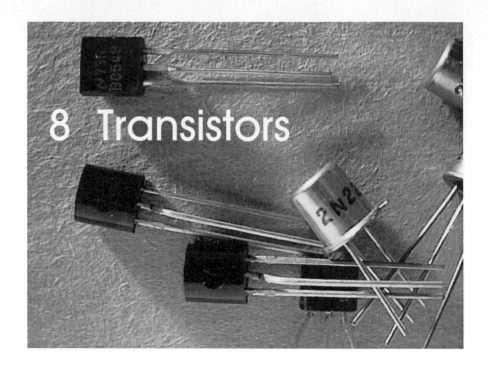

8 Transistors

Transistors of various types form the basis of most electronic circuits. As we shall explain later, there are two fundamental applications for transistors — as electronically controllable switches and as amplifiers. We describe several different types of transistor in this chapter, beginning with the MOSFET field effect transistor, which is the easiest to understand.

MOSFETs

The action of a field effect transistor depends upon the *effect* of the *field* produced by a charged region known as the **gate**. The term MOSFET is the acronym for **metal oxide silicon field effect transistor**. The diagram opposite shows why the transistor has this description. The transistor is based on a bar of p-type *silicon*. Two crosswise strips in the bar are doped to make them into n-type material, and *metal* is deposited on these to form the source and drain terminals. Between the strips, the surface of the silicon is covered with a thin layer of *silicon oxide*, which is a non-conductor. Metal is deposited on the silicon oxide to form the gate electrode.

From the account above we can see that the transistor consists of metal, oxide (of silicon) and silicon. It works by the effect of an electric field.

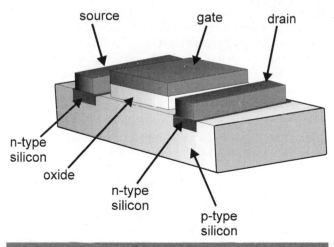

source   gate   drain

n-type
silicon

oxide

n-type
silicon

p-type
silicon

A MOSFET is built on a substrate of p-type silicon.

The transistor is connected into a circuit through its three terminals, source, gate and drain. Some types also have an external connection to the substrate, but in most types this is connected internally to the source as shown below. A potential is applied between the source (0 V) and the drain (positive) but there is a p-n junction between them and the p type material, so no current can flow. It is like having two diodes connected back-to-back. The p-type material is connected to the source so it is at 0 V.

When the gate is made positive of the p-type material, it repels the holes from nearby regions of the p-type material, turning it temporarily into n-type material.

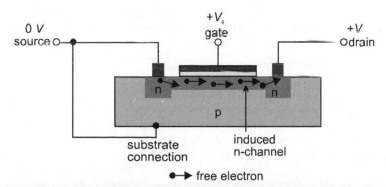

The effect of the field of the gate is to create a channel of n-type silicon connecting the two existing n-type regions.

Another way of looking at this is to say that the gate attracts electrons to the region around it. These are more than enough to fill the holes in that region and the remainder are available as charge carriers. This creates a channel joining the two n-type strips. Current can then flow from the source to the drain. The greater the gate potential, the wider the channel and the larger the current.

The full name of the transistor described above is an n-channel enhancement MOSFET. The term *n-channel* describes the way that charge flows through the transistor, carried by electrons. The term *enhancement* refers to the fact that there is no channel when the gate is at 0 V, but the channel is enhanced (made wider) as the gate potential increases. This kind of MOSFET is the most widely used but there are other kinds, including the p-channel enhancement MOSFET, which is similar to the n-channel type, except that the polarities are reversed.

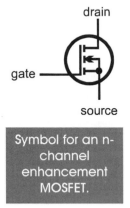

drain

gate

source

Symbol for an n-channel enhancement MOSFET.

Power MOSFETs

MOSFETs built along the lines described, but with heavier construction, may be used for currents of a few amps. Usually such types have a metal tag to which a heat sink (p. 70) may be bolted. Several special types of MOSFET are used for higher currents. One of these types are the **v-channel MOSFETs**, generally referred to as **VMOS**. The 'v' in *v-channel* describes the shape of the channel, not its charge-carriers. In VMOS transistors, the doped layers are built up on a surface that has previously been etched to form steep-sided grooves, v-shaped in section. The action of the transistor is normal. The difference is that conduction is down the sides of the grooves. The distance from top to bottom of the grooves is short, and the channel is very wide in the direction *along* the grooves, so its resistance is low. This means that currents can be large without overheating the transistor. VMOS power transistors are made to carry currents of 10 A or more. They make it possible to control large currents by devices that can produce only a small current. For example, a heavy-duty relay or a 10 Ω loudspeaker can be operated directly from the output of a CMOS logic gate.

Another type of structure is the HEXFET. Like VMOS, it relies on the channel being short but wide to give very low 'ON' resistance and thus allow high currents without overheating.

Gate potential

If the gate of a MOSFET is left unconnected, electric fields from outside can easily charge it. The small charge acquired in this way is enough to produce a significant effect on conduction through the transistor. When a circuit is under construction, there are many ways in which fields can be generated. The gate may be affected even without coming into direct contact with charged surfaces. An unconnected gate is almost certain to lead to erratic behaviour. The rule Is: leave no inputs to gates unconnected.

A problem may arise when MOSFETs are handled. The human body becomes charged due to friction between the body and clothing or floor coverings. Usually, we do not notice this because, although the potential of the body may rise to several hundreds or thousands of volts, the amount of charge is small and the currents produced when we touch grounded objects are usually (though not always) too small to be felt. But, if we touch a gate terminal of a MOSFET, the discharge current can easily penetrate the very thin layer of oxide and destroy the transistor. Precautions must be taken to avoid this happening. These include storing MOSFETs In packages made from conductive foil, or with their terminal wires pushed into conductive plastic foam, not wearing clothing made from synthetic fibres, and wearing an earthed wrist-band.

JFETs

The full name of these is **junction field effect transistors**. This name comes from the fact that its action depends on what happens at a p-n junction (page 74). The structure of a JFET is shown overleaf. A bar of n-type semiconductor has metal contacts at each end. On either side of the bar is a layer of p-type material, the two layers being connected together by a fine wire. Remember that this structure measures only a millimetre or so across.

If a p.d. is applied to the ends of the bar, a current flows along the bar. Because the bar is made from n-type material, the current is carried by electrons. We say that this is an **n-channel JFET**.

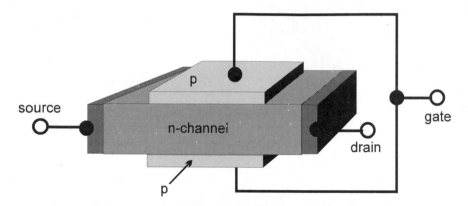

An n-channel JFET has a bar of n-type silicon, sandwiched between two layers of p-type silicon.

When referring to the potentials in this diagram we take the potential of the source to be zero. Electrons enter the bar at the **source** terminal (the reason for its name), and leave at the positive end of the bar, the **drain** terminal. The layers of p-type material are known as the **gate**. If these are at 0 V or a slightly more positive potential, they have no effect on the flow of electrons. But, if the gate is made more negative than the source (perhaps by connecting a cell with its positive terminal to the source and its negative terminal to the gate), an important effect follows. The p-type gate and the n-type bar form a p-n junction. With the gate negative of the bar, the junction is reverse biased and a depletion region (p. 74) is formed. The channel through which they flow is made narrower. In effect, the resistance of the bar is increased.

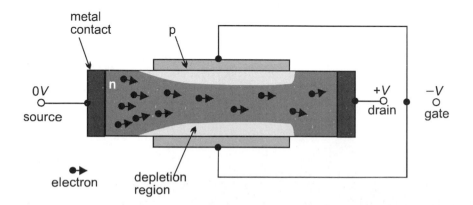

The depletion regions restrict the width of the channel.

The narrowing of the channel reduces the flow of electrons through the bar. There are two ways of using this *field effect* produced by the p-n *junction:*

- **Switching:** If the p.d. between the gate and source is very high, the depletion region extends across the whole width of the bar. This cuts off the flow of electrons altogether. Used in this way the JFET is a **voltage-controlled switch**.

- **Current control:** With smaller gate potentials that do not completely cut off the current, the width of the bar left free for the passage of electrons is proportional to the gate potential. As this is increased, so the width of the channel decreases, the resistance increases and the current decreases. In effect the JFET is a **voltage-controlled resistor**. Putting it another way, a JFET can be used to convert a change of potential to a change of current.

A JFET is always operated with the p-n junction reverse-biased, so current never flows from the gate into the bar. The current flowing into or out of the gate is needed only to change the potential of the gate. Since the gate is extremely small in volume, only a minute current (a few picoamps) is required. A small change in potential of the gate, controls the much larger current flowing through the channel. This property of the transistor can be used in the design of amplifiers.

JFETs have many applications, particularly in the amplification of potentials produced by devices such as microphones which are capable of producing only very small currents. They are also useful in potential-measuring circuits such as are found in digital test-meters (p. 148), since they draw virtually no current and therefore do not affect the potentials that they are measuring (p. 61).

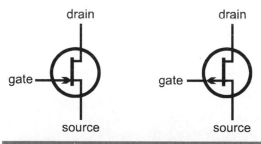

The symbols of (left) n-channel and (right) p-channel junction field effect transistors. Conventional current flows from drain to source.

A JFET similar to that described above is manufactured from a bar of p-type material with n-type gate layers. Current is conducted along the bar by holes and this is known as a p-channel JFET. In operation, the gate is made positive of the source to reverse-bias the p-n junction.

Zero potential

We often mark the terminals of a cell or battery with '–' and '+' to indicate the negative and positive terminals. The negative terminal is the one from which negative charge carriers (electrons) leave the cell. The other terminal is said to be more positive than this, and it is to this terminal that the electrons flow. When we refer to potential, the terms *positive* and *negative* have slightly different meanings. They describe the potential of a point in space or, in more practical terms, of a terminal or an electrode, or any other point in a circuit. In these terms, we can say that one point is 'more positive than' or 'more negative than' some other point. The terms *positive* or *negative* indicate whether the potential goes up or down as we go from one point to another. Potential can only be described relative to the potential at some other point.

From the definition of potential (p.21), zero potential would be the potential of some point in space far away from any charged bodies. Even if such a point exists, it would not be practicable to base our measurements on this. We need a reference point which is closer to hand. Often we take the potential of the Earth as our reference point. We may refer to this as **ground potential** or simply **ground**, and rate this as being 0 V. All potential in circuits connected to ground (that is, earthed circuits) are measured with respect to this.

In circuits that are not actually earthed, it is usually most convenient to take one of the conductors that runs to all or most parts of the circuit as the reference point. Often this conductor is connected to the negative terminal of the cell or other power supply. Although it may not be earthed, this conductor may still be referred to as *ground*. Other points may have potentials positive to this; others may be negative.

In the case of an n-channel JFET, we usually take the source to be at 0 V, the gate to be at negative potential and the drain to be at positive potential.

Bipolar junction transistors

Owing to practical difficulties in making FETs in the early days of semiconductors, bipolar transistors were the first to be widely used. They still are extremely popular with designers, even though the problems of making FETs have been overcome. They work on an entirely different principle to FETs.

A bipolar transistor is a three-layer device consisting either of a layer of p-type sandwiched between two n-type layers or a layer of n-type between two p-type layers. These are referred to as **npn** and **pnp** transistors respectively. The fact that conduction occurs through all three layers, which means that it involves both electrons (negative) and holes (positive) as charge carriers, is why these are sometimes called **bipolar transistors**. Their full name is **bipolar junction transistors** (or **BJT**) because their action depends on the properties of a pn junction (p. 74), as is explained later. The diagram shows the 'sandwich' structure of an npn transistor, the most commonly used type.

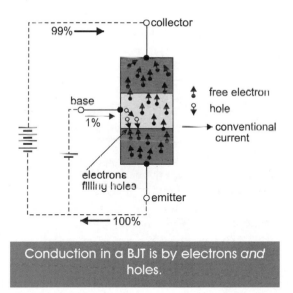

Conduction in a BJT is by electrons *and* holes.

The three layers of the transistor are known as the **collector**, the **base** and the **emitter.** In effect, the transistor consists of two pn junctions (in other words, diodes, connected back-to-back). It would seem that it is impossible for current to flow from the collector to the emitter or from the emitter to the collector.

Whatever the direction of the p.d., one or other of the p-n junctions is sure to be reverse-biased. This is where the features of the base layer are important:

- It is very thin (though not shown like that in the drawing).
- It is lightly doped, so it provides very few holes.

When the transistor is connected as in the diagram, the base-emitter junction is forward-biased. Provided that the base-emitter p.d. is greater than about 0.6 V, a base current flows from base to emitter. When describing the action in this way, we are describing it in terms of conventional current, as indicated by the arrows in the drawing. What *actually* happens is that electrons enter the transistor by emitter terminal and flow to the base-emitter junction. Then, as in a diode (p. 77), they combine with holes that have entered the transistor at the base terminal. As there are few holes in the base region, there are few holes for the electrons to fill. Typically, there is only one hole for every 100 electrons arriving at the base-emitter junction. The remaining 99 electrons, having been accelerated toward the junction by the field between the emitter and base, are able to pass straight through the thin base layer. They also pass through the depletion layer at the base-collector junction, which is reverse-biased. Now they come under the influence of the collector-emitter p.d. The electrons flow on toward the collector terminal, attracted by the much stronger field between emitter and collector. They flow from the collector terminal, forming the collector current, and on toward the battery. In effect, the base-emitter p.d. starts the electrons off on their journey but, once they get to the base-emitter junction, most of them come under the influence of the emitter-collector p.d.

The important result of the action is that the collector current is about 100 times greater than the base current. We say that there is a **current gain** of 100.

If the base-emitter p.d. is less than 0.6 V, the base-emitter junction is reverse-biased too. The depletion region prevents electrons from reaching the junction. The action described above does not take place and there is no collector current. In this sense the transistor acts as a **switch** whereby a large (collector) current can be turned on or off by a much smaller (base) current. If the base-emitter p.d. is a little greater that 0.6 V, and a varying current is supplied to the base, a varying number of electrons arrive at the base-emitter junction. The size of the collector current varies in proportion to the variations in the base current. In this sense the transistor acts as a **current amplifier**. The size of a large current is controlled by the variations in the size of a much smaller current.

The structure and operation of pnp transistors is similar to that of npn transistors, but with polarities reversed. Holes flow through the emitter layer, to be filled at the base-emitter junction by electrons entering through the base.

Symbols for npn and pnp BJTs.

Mass production

Very large numbers of transistors are produced at once on a single silicon wafer. A frequently used method is the **planar** process.

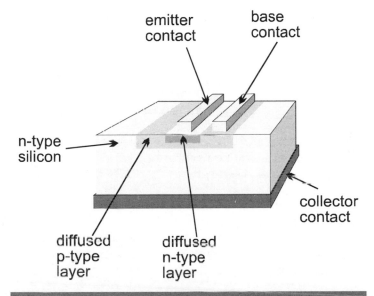

emitter contact

base contact

n-type silicon

collector contact

diffused p-type layer

diffused n-type layer

> In the planar process, successive layers of dopants are diffused into the silicon, using masks.

First a layer of silicon oxide is formed on the upper surface of the wafer by heating it in an atmosphere of oxygen and water vapour. The next step is similar to that used for making pcbs (p. 157), except that it is on a very much smaller scale. The silicon oxide layer is coated with a layer of photoresist. Then the wafer is exposed to uv light, usually through a mask on which the areas to be etched are left clear. The pattern of the mask is repeated over and over again to produce many transistors, which can be later separated by cutting the wafer. Another approach is to have a single mask and to step this along to produce an array of exposed areas. For finely detailed circuits, a reduced image of the mask is projected onto the photoresist layer using a lens system.

Mass production (continued)

The resist is etched chemically, removing the exposed areas of the resist and the oxide beneath. Finally, the wafer is doped, as above, but with a different dopant. This time the dopant can not reach the areas covered with oxide, so only the exposed areas are doped. It is possible to control the depth to which the dopant penetrates.

When an npn transistor, such as that on page 89 is made, we begin with a wafer doped to produce n-type silicon. Areas of this are etched and doped to produce the p-type silicon of the base layer. Although the silicon in the base begins as n-type, it is converted to p-type by diffusing an excess of a hole-producing dopant, to cancel out the effects of the electron-producing dopant already present. The wafer is then exposed to a mask which causes smaller areas to be etched over the base layer, and then the wafer is doped to produce a smaller shallower n-type layer (emitter) in the base layer. Finally the collector (substrate), base and emitter contacts are added by placing the wafer in a vacuum, covering it with a mask and evaporating metal on to it. Alternatively a continuous layer of metal may be deposited and etched away to leave the required connections.

Practical transistors

A transistor is a minute object on a small chip of silicon (rarely germanium). To make it practicable to handle the transistor, it is mounted in a case or sealed into a block of plastic, with thin wires connecting the base, emitter and collector to thicker terminal wires. The title photograph of this chapter shows typical packages used for low-power transistors. These are two of the standard packages used for JFET, MOSFET and BJT transistors.

By varying the amount of doping, the method of doping, and the geometry of the regions, transistors with various characteristics can be made. Some have much higher gain than others (up to 800 times), or may be suitable for operation with high currents. In typical general-purpose transistors, the maximum collector current is only a few hundred milliamps.

This power BJT is rated to pass up to 10 A and run at up to 75 W. It has a metal tag to conduct heat away from the transistor inside. A heat sink is bolted to this to increase the dissipation of heat.

Running at powers up to 115 W, this BJT is housed in a stout metal case that offers a large area of contact to the heat sink. The collector is connected internally to the case, so there are only two terminal wires.

Power transistors are capable of collector currents of up to 90 A. A power transistor is robustly constructed with low-resistance channels to minimize the voltage drop across it. It also has a sturdy metal tag or case for bolting to a heat sink.

Radio-frequency transistors are designed specially for operation at frequencies of several hundred megahertz, and some up to 5 GHz. At high frequencies, the capacitance between the base and emitter of a BJT may act to reduce the amplitude of the signal, so radio-frequency transistors are designed to minimize this effect. A transistor designed to operate at radio frequencies may be used in circuits other than radio transmitters and receivers. Many other kinds of device such as computers, mobile telephones, digital cameras and CD players operate at radio frequencies and high frequency transistors are required for these too. Such devices are digital rather than analogue and the main function of the transistors is high speed switching. As explained in the next chapter, the gate or base of a transistor is biased ready for action by connecting it through a resistor to the positive supply lines. During manufacture, there is no problem (and almost no extra cost) in putting the biasing resistors on the same chip as the transistor. This simplifies the layout of the circuit board and saves the cost of separate resistors. **Digital transistors** with resistors included are often used for switching in digital circuits.

Most transistors are available also as **surface mount transistors**. The typical package, measuring only 3.0 mm × 1.5 mm is shown at top centre in the photograph on page 166.

Darlington transistors

A Darlington pair consists of two npn transistors connected as shown in the diagram. The way they work is easier to understand if it is explained in terms of conventional current. Given a small base current flowing to Q1, a much larger collector current flows into the transistor and out of the emitter. This becomes the base current of Q2, which amplifies the current still further. For small base currents, the gain of the Darlington pair equals the gain of Q1 multiplied by the gain of Q2. A typical Darlington pair has a gain of 10 000, and some have gains up to 50 000.

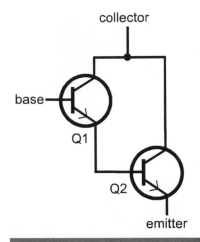

In a Darlington pair, the emitter current of Q1 is the base current of Q2.

A Darlington pair may be assembled from two individual transistors but it is more convenient to use a **Darlington transistor**. This consists of two npn transistors made as one unit with internal connections and enclosed in one of the standard transistor packages. Paired FETs are available, often under the name **FETlington transistors**.

Thermionic valves

Although this is a chapter on transistors, which are solid state devices (that is, based on semiconduction), we should mention one of their vacuum tube equivalents. Such devices, once the only active devices available, have been almost entirely replaced by semiconductors. They still have applications in high-power circuits, such as radio transmitters, and there are audio enthusiasts who claim that the reproduction of music by a valve amplifier gives a tone far more pleasing than that from a solid-state amplifier. There are also the vintage radio experts who take a pride in collecting and restoring long defunct radio equipment and getting it to work again. Valves are still being manufactured.

The simplest thermionic valve is the diode which we have already mentioned (p. 77). This is the functional equivalent of the p-n semiconductor diode. Like all thermionic devices, it operates with a high cathode-anode voltage, usually over 100 V. The next most complex valve is the **triode,** which, as its name implies comprises *three* electrodes.

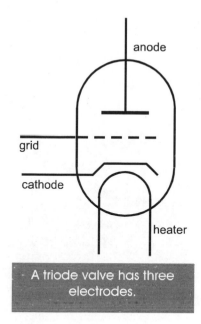

anode

grid

cathode

heater

A triode valve has three electrodes.

The symbol for a triode shows its basic structure but, in practice, the electrodes are differently arranged. In the centre of the tube is the heater filament, wound on an insulating former. Close around this is the cathode, a cylinder of thin metal. This may be connected directly to the heater, as in the diode diagram on page 77, or the cathode may have a separate terminal, as on the left. The next electrode is a cylinder around the cathode and concentric with it. This is made from wire mesh and is called the **grid**. On the outside, concentric with the other two cylinders is the anode, made from thin metal. The whole structure is enclosed in an evacuated glass tube.

In action, a current is passed through the heater, which glows red hot and heats the cathode. The heated metal atoms liberate free electrons which form a cloud around the cathode. The cathode is at 0 V, and the anode at a high voltage, 100 V or more. This causes the electrons to be attracted toward the anode. Current flows through the valve. So far, the action is the same as we have described for a diode. Electrons can pass through the meshes of the grid but, if the grid is made negative of the cathode, some of them are repelled and travel back toward the cathode. The more negative the grid, the more electrons are repelled. In other words, the flow of electrons, and hence the current through the tube, is controlled by the *field* around the grid and, hence, by its potential. In this way the triode is the equivalent of a field effect transistor. The fact that small changes in grid potential can cause large changes in anode-cathode current is the basis of many kinds of amplifying circuit.

Many other types of thermionic valve are made, some with four, five or more electrodes to give superior performance.

9 Semiconductor circuits

This chapter describes some of the ways in which transistors and diodes are used.

A MOSFET switch

The action of a MOSFET when used as a switch can be demonstrated by a simple circuit. The circuit is supplied from a 6 V battery and a variable voltage is obtained from a variable resistor VR1, acting as a potential divider. The potential at the wiper of VR1 is fed to the gate of an n-channel MOSFET, Q1. In this way we can set the gate-source p.d., V_{GS}, to any value between 0 V and 6 V. We want to know what effect this has on the drain current, I_D.

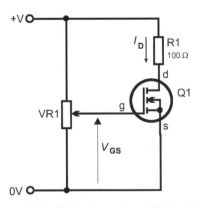

In this simple switching circuit, the current through the MOSFET is controlled by its gate-source p.d.

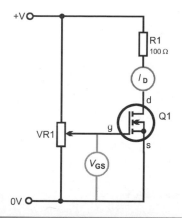

The source of the transistor is connected to the 0 V line. The drain is connected through a resistor R1 to the positive supply line. The resistor is known as the **load**. It could be anything with a resistance of 100 Ω that is switched on and off by the transistor, such as a lamp, a buzzer, or an electric motor.

To find out what happens, we need to connnect two meters to the circuit. The voltmeter measures the potential (in volts) being applied to the gate. The milliammeter measures the current flowing through the load.

A voltmeter and milliammeter are connected as shown to investigate the behaviour of the MOSFET.

Refer to the stuctural drawings of a MOSFET (p. 81). When the gate is at 0 V, there is no channel between the two n-type regions. Therefore there is no current through the transistor. As we turn the knob of VR1, the gate-source p.d. increases steadily. As it reaches the threshold voltage (about 1.95 V for this transistor) the channel has become wide enough for a small current to flow. As the gate-source p.d. increases above the threshold, the channel becomes wider still. The current increases to about 60 mA.

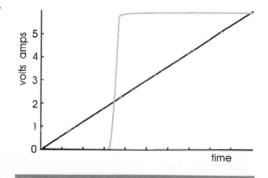

The voltage at the gate (black) increases as the knob of VR1 is turned. The current (grey) is zero at first but increases as V_{GS} passes the threshold voltage.

When the gate-source p.d. has reached about 2.2 V, the channel has reached its maximum width and is conducting so well that it has a resistance of only a few ohms. The transistor is now fully switched on. We say that it is **saturated**. Further increase of gate potential has almost no effect. Note that the input to the MOSFET is a *p.d.* (or voltage) and its output is a *current*.

Measuring potential difference and current

P.D. (or voltage) is measured with a voltmeter. The meter is connected to the two points in the circuit for which the p.d. is being measured. A voltmeter has *high* resistance, so very little current is short-circuited through the voltmeter.

Current is measured with an ammeter. It is connected into the circuit so that the current flowing in the circuit flows through the ammeter. An ammeter has low resistance, so it does not restrict the flow of the current that is being measured.

A MOSFET amplifier

As can be seen in the graph on the previous page, there is a range of gate-source p.d.s in which the transistor is switched partly on, but is not saturated. If we operate in this range we can use the transistor as an amplifier. The first thing to do is to bias the gate into the operating region. The diagram shows how this is done, using a potential divider made from two fixed resistors.

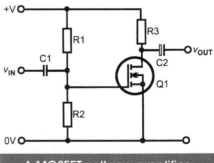

A MOSFET voltage amplifier.

The signal, v_{IN}, is fed to the transistor by way of a coupling capacitor C1 (see p. 46). The signal source could alternatively be connected directly to the gate but capacitor coupling allows the average voltages on either side of the capacitor to be unequal. The source might be a crystal microphone with an average (no sound) output voltage of 0 V. Sound would make its output oscillate a few millivolts positive and negative of zero. On the other side of C1, the resistors fix the average voltage at, say 2V. When a signal arrives from the microphone the voltage at the gate oscillates a few millivolts positive and negative of 2 V. This causes the width of the channel to vary slightly, but not enough to turn the transistor off, or to saturate it. In response to this, the current through Q1 and R3 varies in size. The transistor has converted a voltage of varying size into a current of varying size.

Note that with capacitor coupling it is essential for the signal to be an alternating one, such as an audio or video signal. Direct current (that is, a constant voltage level) does not pass through a capacitor.

To be an amplifier, both input and output must be *voltages*. The output v_{OUT} must have greater amplitude than v_{IN}. The function of R3 is to convert the varying current through Q1 and R3 into a varying output voltage. The p.d. across a resistance is proportional to the current (p. 25). In this circuit, one end of the resistor is connected to the positive supply. If there is only a small current through R3, only a small p.d develops across R3. The other end is only slightly below the supply voltage. On the other hand, if there is a large current through R3, there is a large p.d. across it, and the voltage at its other end (where it connects to Q1) drops to a low value. The voltage at this junction varies with the input signal and a capacitor C2 couples the amplifier output to the next stage. This might be a further amplifier or a headset.

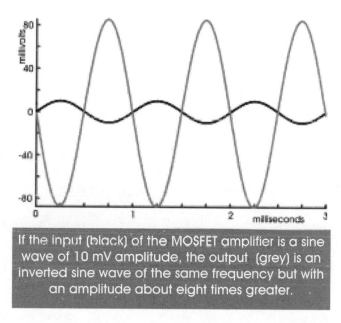

If the input (black) of the MOSFET amplifier is a sine wave of 10 mV amplitude, the output (grey) is an inverted sine wave of the same frequency but with an amplitude about eight times greater.

The graph plots the input and output voltage signals of the MOSFET amplifier, when a 1 kHz signal is fed to it. Depending on the value of R3, the voltage gain of the amplifier is approximately −8. The negative sign indicates that the output signal is inverted with respect to the input signal.

In this amplifier, the source terminal is common to both the input and output sides of the circuit. For this reason this type of amplifier is known as a **common source amplifier**.

Transistor connections

There are three basic ways in which a transistor may be connected and used in a circuit. In the previous section, we saw a MOSFET used in the common source connection. The equivalent connection for an npn BJT is the **common emitter** connection. The other connections for a BJT are the **common base** and **common collector** connections.

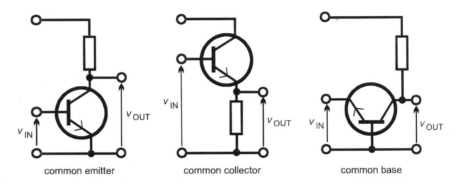

common emitter   common collector   common base

The three basic ways of connecting a BJT. Equivalent connections are used for MOSFETs and JFETs.

The common emitter connection is the most frequently used. It corresponds to the common source connection used for FETs. Essentially and unlike an FET, a BJT is a *current amplifier*. A small base current controls the flow of a much larger collector current. If we apply a varying voltage to the base, usually through a resistor to limit the size of the base current, there is zero collector current until the base-emitter voltage rises above 0.6 V. At this point it overcomes the virtual cell of the base-emitter p-n junction (p. 76). In a BJT amplifier, the base is biased by one or two resistors to keep it a little way above 0.6 V. The main difference between the amplifier circuit for a BJT and the equivalent circuit for an FET is that virtually no current flows into the gate of an FET. This is because a typical gate has an input impedance of 10^{12} Ω. Currents will be of the order of millionths of a microamp. By comparison, the base current of a BJT is usually measurable in milliamps. The collector current is about 100 times larger than this (p. 87) and flows through a collector resistor, the equivalent of R3 in the MOSFET amplifier. As in that amplifier, this produces a varying p.d which gives rise to a varying output voltage.

As in the common source amplifier, the output signal of a BJT common emitter amplifier is amplified and inverted.

Common collector amplifiers

It is clear from the simple diagram of this amplifier on the opposite page, that there must always be a constant voltage difference between v_{IN} and v_{OUT}. This is the 0.6 V 'voltage drop' across the p-n base-emitter junction. Whatever the value taken by v_{IN}, the value of v_{OUT} is always 0.6 V less. The emitter potential follows the input, so a common collector amplifier is also known as an **emitter follower** amplifier.

As an amplifier, this circuit has a voltage gain of one. There may not seem much point in having an amplifier with a gain of one, but the circuit has its uses. Like all base currents, the base current is very small. The circuit takes very little current to drive it. It can be driven from another circuit which is able to supply only a very small current. An example is a crystal microphone (p. 118) which produces current of only a few microamps. But the common emitter amplifier is able to *supply* a very much larger current, its collector current, to anything connected to its output terminal. It has a high **current gain**, often greater than 100.

Note that, as input potential goes up, so does output potential. Unlike the common emitter amplifier, this amplifier is **non-inverting**.

It was explained on page 61 that the voltage output from a potential divider may fall significantly if too much current is drawn from it. An emitter follower amplifier may be used to remedy this.

The emitter follower draws only a small current from the potential divider, yet is able to supply a large current to the load. The divider is set to a voltage 0.6 V higher than that needed by the load.

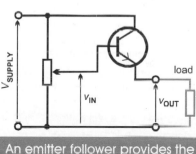

An emitter follower provides the load with ample current without overloading the potential divider.

Common base amplifier

The third type of transistor amplifier has the base of the transistor common to both sides. The input is fed to the emitter and the non-inverted output is taken from the collector.

A common base amplifier has a high voltage gain, as high as that as the common emitter amplifier, but there are problems when we consider current. Its input requirements are high and its current output is low.

It might seem that this type of amplifier has little to recommend it, but it has one feature that is useful for certain applications. One of the disadvantages of the common emitter amplifier is that it is subject to the **Miller Effect**. At high frequencies there is an effect due to the capacitance between the base and collector layers of the transistor. This is made worse by the fact that the current on the collector side is 100 times higher than that through the base. This greatly increases the effective capacitance and there is a reduction of signal strength at high-frequency signals.

In a common base amplifier, the signal enters through the emitter and leaves through the collector. The emitter and collector are separated from each other by the base layer, so the capacitance between them is less. This makes common base amplifiers specially suitable for high-frequency signals. These are often used in aerial preamplifiers for UHF television.

Protective diode

A precaution must be taken when using a transistor to switch an inductive load, such as a relay, a solenoid or a motor. When the transistor switches off, the current through the load ceases instantly and the magnetic field in the coil collapses. The effects of self induction cause a very large e.m.f. to be generated (p. 51), so as to attempt to make the current continue to flow. This causes a p.d. to be produced across the load which is very much higher than the p.d. originally present. It could easily damage the transistor.

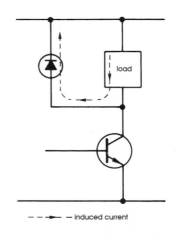

- - - ►- — induced current

Protecting the transistor.

To prevent this we connect a diode across the load. Current does not normally flow through the diode, as it is reverse-biased. But, when self induction causes the large current (dashed line) to flow down through the load from the positive rail, the diode diverts this current back to the positive rail, and it does not flow into the transistor and damage it.

Schmitt trigger

One of the drawbacks of using a single transistor as a common emitter switch is that it switches on and off gradually. If the input voltage is one that changes slowly, such as the voltage from a light sensor circuit, the switch may spend several minutes in an intermediate state. Enough base current is flowing to produce a collector current but not enough to fully saturate the transistor. So the load is switched half on and may waver in its action, especially if the light level is fluctuating slightly. We need a switching circuit that has a more definite 'snap' action. This is provided by the Schmitt trigger circuit.

This description of the way this circuit works begins with the input (v_{IN}) at a potential less than 0.6 V. Transistor Q1 is off, so its collector potential at A is almost equal to +V. A small current flows through R1 and R2 to the base of Q2, turning it on. The large collector current of Q2 flows through the load and energizes it. The same current flows through R3. Since a current is flowing through R3, there is a small positive potential at point B. For the sake of the discussion, assume that this is 1 V.

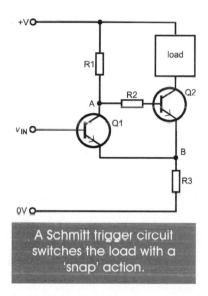

A Schmitt trigger circuit switches the load with a 'snap' action.

If the input potential is gradually increased, nothing happens until the base-emitter p.d. of Q1 exceeds 0.6 V and a base current begins to flow. But, since point B (and therefore the emitter of Q1) is already at 1 V, the base of Q1 must be raised to 1.6 V before Q1 begins to turn on. When Q1 begins to turn on, the potential at A begins to fall. This reduces the base current of Q2. Its collector current is reduced. The current through R3 is reduced and so is the p.d. across it. The potential at B begins to fall.

As the potential at B falls, so the base-emitter p.d. of Q1 begins to increase. This is because the base is at 1.6 V but the potential at B is now less than 1 V. This increasing base-emitter p.d. turns Q1 on even more strongly. Its collector current increases further, decreasing the potential at A and turning Q2 further off. Once begun, the action continues more and more rapidly until Q1 is fully saturated and Q2 is fully off. The load is switched off.

It can be seen that only a very small rise of input above 1.6 V is enough to trigger off the events described above. However, a correspondingly small fall of input below 1.6 V does not have the reverse effect. Now that Q2 is off, there is no current through R3 and point B is at zero. To make the base-emitter p.d. of Q1 less than 0.6 V, the input must be brought below 0.6 V, not 1.6 V. The input has to fall by 1 V or possibly more. When this happens, the reverse action occurs, Q1 turns off and Q2 and the load is switched on again.

To summarize the action, the circuit is triggered when the input increases slightly above the **upper threshold** (1.6 V in the example). Once triggered, it is not triggered in the reverse direction until the input has fallen below the **lower threshold** (0.6 V in the example). One feature of the action is that only a slight increase or decrease at the appropriate threshold is enough to initiate an abrupt change in the state of the circuit. The other feature is that the thresholds are relatively far apart. The exact values of the thresholds and their distance apart may be set by choosing suitable values for the resistors. The difference between the upper and lower thresholds is called the **hysteresis** of the trigger.

10 Power supply circuits

Although cells and batteries are a handy way of supplying power to portable equipment, they are expensive to replace. Even rechargeable cells have a limited life. For many kinds of equipment, it is preferable to obtain power from the mains and convert it to a form suitable for driving electronic circuits. This chapter deals with such power supplies.

Rectifiers

Almost all electronic circuits need a d.c. power supply. The action of an npn transistor requires the collector to be positive of the emitter, and it does not work if the polarity is reversed. An alternating power supply is of no use. The same applies to other transistors and related devices. Batteries provide a d.c. supply and are particularly suitable for portable equipment, but, for fixed equipment, it is more economical in the long run to use the alternating supply provided by the mains. This has far too high a voltage (230 V) for use with most electronic circuits. This is no problem because, since the current is alternating, it can be stepped down to a suitable level with a transformer (p. 55). The output of the transformer is a low-voltage alternating supply. The next requirement is to convert the a.c. supply into a d.c. supply. This process is known as **rectification** and a circuit which does this is called a **rectifier**.

Single-diode rectifier

The obvious choice for a rectifying device is a diode, for this allows current to flow through it in only one direction. Any diode can be used, but special rectifier diodes are made, suitable for withstanding the large currents and reverse p.d.s usually found in rectifier circuits.

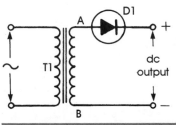

On the left of the diagram is a step-down transformer rated to produce the required voltage from the mains supply. On the right is a single diode which performs the rectification. The waveform of the a.c. from the transformer has the shape of a sine wave. If it is taken from the mains supply, it has a frequency of 50 Hz. One complete wave takes a fiftieth of a second so the total time for the three cycles shown in the graphs below is three-fiftieths of a second or 0.06 s.

A single diode functions as a half-wave rectifier , but this is only 50% efficient.

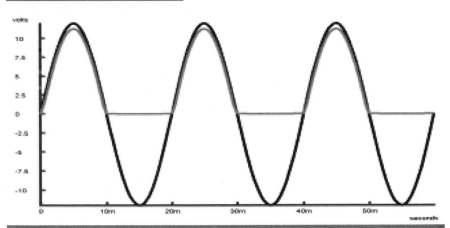

The input (black) to the half-wave rectifier is AC at 50 Hz. The output is pulsed DC (grey). The diode conducts on the positive half-cycle.

During the first half of each cycle the waveform is positive. This means that current flows out of terminal A of the transformer, through the diode, through any circuit that happens to be connected, and back into the transformer coil through terminal B. During the second half of the cycle, current would flow out of B and into A but this is prevented because of the diode.

The result is shown by the grey curve. The diode conducts during the first half of each cycle and a pulse of current flows around the circuit. No current flows during the second half of the cycle.

This type of rectifier produces one-way current but has the disadvantage that it supplies current for only half the time, so it is inefficient. It is known as a **half-wave rectifier**. The graph shows that the maximum output voltage is 0.6 V less than the input. This is due to the voltage drop across the diode

Bridge rectifier

A diode bridge consists of four diodes connected as in the drawing. This circuit produces current during both halves of the cycle.

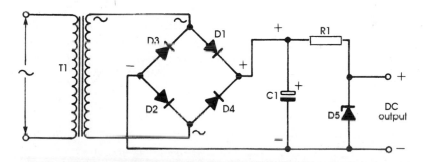

From left to right, this power supply circuit comprises a transformer, a rectifying bridge, a smoothing capacitor and a Zener voltage regulator.

During the first half of the cycle, current flows from the transformer (from the upper terminal in the diagram), through diode D1, round the circuit, through diode D2, and finally back to the transformer. The diodes between D1 and D2 are conducting, but the other two diodes are reverse-biased so they do not conduct. During the second half of the cycle current flows from the transformer (from the lower terminal in the diagram), through D4, round the circuit in the same direction as before, through D3 and finally back to the transformer. The diodes D3 and D4 are conducting, but the other two diodes are reverse-biased so do not conduct. This produces the waveform shown in the graph overleaf, and the diode bridge is known as a **full-wave rectifier**. A bridge rectifier may be assembled by connecting four individual diodes, or a rectifier bridge may be used. This unit consists of four diodes ready-connected and sealed into a single four-terminal package.

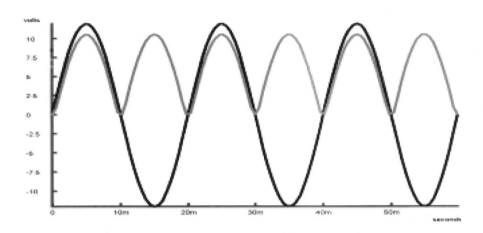

Without the electrolytic capacitor shown in the diagram on page 105, the output (grey) waveforms of a full-wave rectifier is pulsed DC with a frequency double that of the input (black). The amplitude is 1.2 V less than the input because of the voltage drop across *two* diodes.

Smoothing

The output of half-wave and full-wave rectifiers is pulsed d.c., quite unlike that obtained from a battery. It is suitable for powering some kinds of circuit, but not for all. If a pulsed supply is used with an audio circuit, for example, the result is a strong 'mains hum' at 100 Hz (double the mains frequency) superimposed on the audio signal. Similarly, a pulsed supply is unsuitable for powering logic circuits.

The solution is to connect a large-capacitance capacitor across the output from the rectifier. Capacitances of several thousand microfarads are required so we normally use an electrolytic capacitor, as can be seen in the title photograph of this chapter. During the peak of each half-cycle the capacitor receives a boost of charge from the rectifier. The charge passes out to the powered circuit at a steady rate so there is a slight fall of voltage after each peak. If the capacitance is sufficiently large, voltage will have fallen by only a small amount before the capacitor is recharged by the next peak. As shown in the graph opposite, the waveform of smoothed d.c. is a more-or-less steady voltage with a slight ripple at 100 Hz.

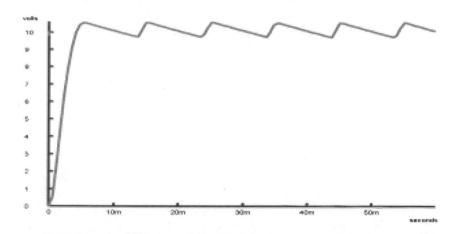

Adding a capacitor to the output side of the rectifier produces D.C with a slight 100 Hz ripple.

Voltage regulation

Mains power supply units (often made so as to plug into an ordinary mains socket) are sometimes sold as 'battery eliminators'. They are used to provide power for running a radio set or tape recorder, and are more economical than using batteries. Their output is d.c. and may be rated at 6 V, 9 V and sometimes other set voltages. If you measure the output using a test-meter, you may find that the voltage is much higher than the rated voltage when only the meter is connected to the unit. When a radio set or other piece of equipment is connected, the output falls to the rated level, but may possibly be several volts lower. This is because the voltage of a simple transformer-rectifier-smoother falls as the current drawn from it increases. This does not usually matter for radio sets and similar equipment but it does matter if parts of a circuit are designed to operate at a fixed voltage.

If it is necessary to regulate the voltage from a power supply, an additional stage is required. In the diagram on page 105, the additional stage is made up from a resistor and a Zener diode. The Zener diode is chosen to have a Zener voltage (p. 79) equal to that required by the circuit which is to be supplied. This must be lower than that provided by the rectifier circuit. When current is flowing to the external circuit there is a p.d. across the resistor. The value of this is chosen so as to bring the voltage down to slightly more than that required by the external circuit when it is using its maximum current.

The zener conducts a small surplus current and the remainder goes to the external circuit. The p.d. across the d.c. output terminals is the Zener voltage. If the external circuit changes its requirements so that it needs less current, the excess current flows away through the diode. The supply to the external circuit remains at the Zener voltage. Regulation by a Zener diode is not perfect but is adequate for many purposes.

A **bandgap voltage reference** has a similar action to that of the Zener diode. On page 116, it is explained how a bandgap device may be used as a temperature sensor, by adjusting the rates of change of two opposing p.d.s. In the bandgap voltage reference the adjustments are such that the same voltage difference is obtained at all temperatures within a wide range. The reference therefore gives a constant voltage at any temperature, making it suitable for precision circuits. It can replace the Zener diode on page 105, to produce a better degree of regulation of the output voltage.

An even better way to regulate output voltage is to use an emitter follower amplifier (p. 99) as a voltage regulator. On the unregulated side of the circuit (after the smoothing capacitor), current flows through a resistor and a Zener diode. The diode is reverse biased into the avalanche breakdown region (p. 79). The value of the resistor is such that the reverse current is quite small, say 5 mA.

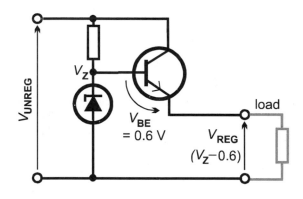

V_{UNREG} varies as the current being drawn from the circuit rises and falls. The current flowing through the diode increases above 5 mA, or decreases below 5 mA, but the p.d. across the diode remains close to the Zener voltage. In this way, the Zener diode holds the base of the transistor at its Zener voltage, V_Z.

The emitter follower amplifier has applications as a voltage regulator.

Variations in the amount of current drawn from the circuit (within limits) have no effect on the voltage at the base of the transistor. The transistor is usually rated to pass currents of 1 A, possibly more. Current flows through the collector and emitter to the external circuit.

As long as the transistor is conducting, there is the usual p.d. (V_{BE}) of about 0.6 V between the base and the emitter. This is due to the virtual cell at the p-n junction. So the regulated output voltage V_{REG} is 0.6 V less than V_Z. For example, if the Zener voltage is 4.7 V, the regulated output voltage is 4.1 V.

A transistor circuit such as the above may be incorporated into a type of integrated circuit (p. 159) known as a **voltage regulator**. Such devices may also have current limiting features. If the load develops a short circuit or if the current drawn exceeds a safe amount for any other reason, this condition is detected by the regulator and the voltage output is sharply reduced. The circuit may also include a thermistor circuit (p. 115), which provides thermal shutdown, cutting off the current if the device becomes overheated. Voltage regulators are made to provide a single fixed voltage in a standard range, including 5 V, 6 V and 12 V, as well as a range of negative voltages. Variable regulators are also available.

Silicon controlled switches

Members of this family of devices are variously known as silicon controlled switches (SCSs), silicon controlled rectifiers (SCRs) or **thyristors**. These are four-layer devices consisting of alternate n-type and p-type layers.

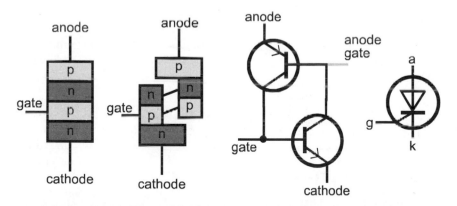

A thyristor is best thought of as an npn BJT connected to a pnp BJT.

The device is connected with the anode (the emitter of the pnp transistor) more positive than the cathode (the emitter of the npn transistor). The third connection is to the base of the npn transistor and the collector of the pnp transistor, and is named simply the gate, or the cathode gate.

To begin the cycle of operation, the gate is at, or close to, the potential of the cathode, so the npn transistor is switched off. No current is being drawn from the base of the pnp transistor, so this is switched off too.

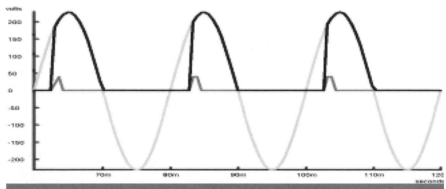

A small positive pulse turns on a thyristor for the remainder of the positive half-cycle.

The npn transistor is turned on when a short positive pulse is applied to the gate. This makes the potential at the collector of the npn transistor fall, drawing current from the base of the pnp transistor. This turns the pnp transistor on. Current flows from its collector. Even though the pulse originally applied to the gate may have finished, the current from the pnp transistor keeps the npn transistor switched on. As a result, current continues to flow through the device for as long as there is a positive potential at the anode. The device has been switched on by a very brief pulse at its gate and, once switched on, remains on.

pnp transistor action

In the diagram of a pnp transistor, the arrowed line is the emitter. This is usually connected to the positive rail of the circuit. No current flows as long as the gate is held at, or close to, the same positive potential. But, if the gate potential drops 0.6 V or more below the positive potential, a small current flows out of the gate, causing a much larger current to flow through the transistor and out of the collector.

If there is a break in the circuit that prevents current flowing to the anode, both transistors are switched off. They cannot be turned on again by reconnecting the supply to the anode. The only way to turn them on is to reconnect the supply and *then* apply another positive pulse to the gate. The device may also be turned off by reducing the anode-cathode p.d. This reduces the current through the device, but it remains on until the current reaches a minimum value, known as the **holding current**. The device is also turned off by reversing the anode-cathode p.d., making the cathode positive with respect to the anode. In this way, the device acts as a diode, allowing only one-way current, as its symbol suggests. But, unlike a diode, it does not turn on again when the p.d. is restored to its original direction.

Some thyristors are supplied with a fourth terminal, the anode gate. A positive pulse to this switches off the pnp transistor, cutting off the current even though the anode remains positive to the cathode.

A thyristor earns the name of *silicon controlled rectifier* in circuits that make use of its rectifying properties. Its big advantage is that the rectifier is under electronic control.

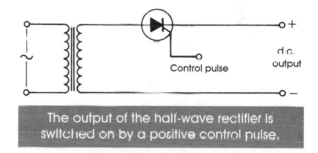

The output of the half-wave rectifier is switched on by a positive control pulse.

A triggering circuit produces a short positive pulse at the same stage during each positive half-cycle of the supply. When triggered, the thyristor behaves like the diode of a half-wave rectifier (p. 104). But, as a.c. voltage falls near to zero, the current dies away and eventually falls below the holding current. Conduction stops and does not start again until the thyristor is triggered during the first half of the next cycle. The action can be seen in the diagram opposite. The triggering pulse is timed to occur about one third of the way through the positive half-cycle. Immediately, the thyristor begins to conduct and the waveform is a sine wave until the a.c. voltage falls to zero and goes negative. It does not conduct again until it is triggered again during next positive half-cycle. As a result, current flows for only one third of a cycle. By altering the timing of the trigger pulse we can arrange for conduction to commence earlier or later in the cycle. In this way the average current flowing to the external circuit can vary between the maximum (equivalent to half-wave d.c.) and zero. Circuits of this type are in common use for controlling the brightness of room lights and the speed of motors on equipment such as electric drills. In the case of equipment operating at mains voltage there is no need for the transformer.

Triac

This is virtually a two-way thyristor, capable of conducting in either direction. Its structural diagram shows it to consist of two thyristors connected in parallel in the opposite sense and sharing a common gate. A positive pulse to the gate causes it to conduct. Like a thyristor, it switches off if the p.d. is reversed but, unlike a thyristor, it can be triggered into conducting in the opposite direction by another positive pulse to its gate.

In a power control circuit, the thyristor shares the disadvantage of the single diode, that it can only conduct for a maximum of half the time. The triac is the solution to this problem. It can be triggered in each half of the cycle, so the power obtainable can be varied between zero and that of full-wave d.c.

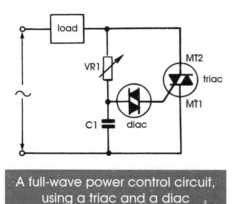

A full-wave power control circuit, using a triac and a diac

In a simple triggering circuit, the alternating potentials in the circuit result in rapidly changing potentials between one plate of capacitor C1 and the variable resistor VR1. By altering the setting of VR1, the timing of the charging and discharging of C1 may be adjusted. This allows trigger pulses to be generated at any required stage during the a.c. cycle.

Triggering is effected by another semiconductor device, a **diac** (see next section). When the p.d. across it exceeds its breakdown voltage, the avalanche effect (p. 77) produces a sharp positive pulse which triggers the triac.

Note that this is not a rectifier circuit. When the triac is in its 'on' state, the d.c. current flows through the load and the triac in alternate directions. The load may be a lamp, a motor, or some other device that is able to work with alternating current.

Diac

The symbol of this device indicates that it has the action of a triac, but lacks the gate. It can be thought of as two diodes connected in parallel, in opposite sense. This gives it its alternative name, **bi-directional trigger diode**.

The diac does not conduct if the p.d. between its terminals is less than a given amount, the *breakdown voltage*. The diodes are of the avalanche type so that an appreciable current flows if the breakdown voltage is exceeded.

IGBT

These are a relatively recent development that combine MOS and BJT technology in a single four-layered device. An IGBT can be described as a BJT that, instead of the base layer, has an insulated gate similar to that found in MOSFETs. This gives them their name **insulated gate bipolar transistors**.

From the functional point of view, the IGBT can be looked on as a power diode that can be turned on or off by the voltage applied to its gate. The advantage of the insulated gate is that they can be turned on or off at high speed by suitable voltage levels. Most of them operate at 10 kHz or more and the faster types operate at up to 200 kHz. They combine this advantage with the high voltage capability of BJTs. Some operate at over 1000 V. These features give IGBTs wide application in switched-mode power supplies.

The detailed circuitry of **switched-mode** supplies is outside the scope of this book. Briefly, instead of regulating the output of the rectifier by a linear circuit such as on page 108, the output is switched on and off at a high rate (10 kHz or more). The ratio between on time and off-time determines the level of the output voltage. Such an output requires heavy smoothing. The advantage of this type of supply is that the output transistor (or IGBT) it always fully on or fully off. It dissipates very little energy in both of these states, and so does not become excessively hot. This contrasts with the linear regulator in which the output transistor is always partly on. It always has an appreciable voltage across it and so dissipates (and wastes) considerable amounts of power ($P = IV$, see p. 33). A disadvantage of switched-mode power supplies is that the supply carries a high frequency component that is the result of the switching.

11 Sensors and transducers

Although the words 'sensor' and 'transducer' are often used as if they have the same meaning, their purpose is entirely different. A **sensor** is a device intended for detecting or measuring a physical quantity such as light, sound, pressure, temperature or magnetic field strength. It has an electrical output (often a varying potential or a current) which varies according to variations in the quantity it is detecting. A **transducer** is simply a device for converting one form of energy into another form of energy.

Some sensors are also transducers. For example, a crystal microphone (p. 118) converts the energy of sound waves into electrical potentials. It conveniently converts the sound energy into a form that can be handled by an electronic circuit. But some other sensors are not transducers. For example, a platinum resistance thermometer detects change in temperature as a change in the electrical resistance of its sensing element. Resistance is not a form of energy. Conversely, there are many transducers that are not sensors; examples include electric lamps (convert electrical energy to light energy), electric motors (convert electrical energy to motion), cells (convert chemical energy to electrical energy).

Descriptions of transducers begin on page 125. Light sensors and transducers are dealt with separately in Chapter 12.

Temperature sensors

The resistance of metals increases with increasing temperature. If we measure the resistance of wire of known length, diameter and composition, we can determine its temperature. A **platinum resistance thermometer** consists of a coil of platinum wire wound on a ceramic former. One of its advantages is that it can be used over a very wide range of temperatures, from –100°C to several hundred degrees Celsius. The resistance of platinum, like that of all metals is very low, so that a long wire is needed for the coil, which is bulky. This means that it takes a relatively long time to acquire the temperature of the surroundings, and also that it is unsuitable for use in measuring the temperature in confined spaces. Finally, the change in resistance is slight and it requires special circuits (p. 62) to measure it. It is chiefly used for high-precision measurements in industrial and scientific applications.

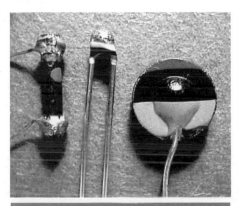

Thermistors are made in the form of rods, beads or discs. The bead thermistor above is only 2 mm in diameter.

A **thermistor**, or thermally sensitive resistor, is much less expensive and easier to use. It consists of a sintered mixture of sulphides, selenides or oxides of nickel, manganese, copper, cobalt, iron or uranium. It is formed into a rod or bead or a disc, and may be encapsulated in glass. The resistance of a thermistor decreases with temperature (we say that it has **negative temperature coefficient, ntc**) and the change is relatively greater than for metals. This makes it easier to detect and measure small temperature changes.

The small size of thermistors makes them suitable for measuring the temperature inside small spaces. For example, they have been used to measure temperatures inside the leaf of a plant. Their small size also means that they have very little effect on the temperature of the objects or spaces into which they are inserted, which makes their measurements more accurate. But thermistors have some disadvantages too. One is that a current must be passed through the thermistor in order to measure its resistance. This inevitably generates a certain amount of heat, which leads to errors in the readings. Measuring circuits must be designed to minimize the current through the thermistor.

The other problem is that the response of a thermistor is not linear. In other words, we do not obtain a straight line when we plot resistance against temperature. This is not significant over a small temperature range but reduces precision over a wider range.

Thermistors are also made with **positive temperature coefficient**. This means that their resistance increases with increasing temperature. Such thermistors are often used for temperature compensation in electronic circuits. Currents in parts of a circuit may increase to dangerously high levels if the temperature of the circuit becomes unduly high (see thermal runaway, p. 70). If a ptc thermistor is included in the circuit, its increased resistance reduces the current.

A **bandgap sensor** depends upon two opposite changes of p.d. The base-emitter p.d. of a transistor decreases with increase of temperature. It is also possible to connect two transistors so that they produce a p.d. that increases with temperature. By suitable choice of resistors it is possible to design a circuit in which these counter-balancing p.d.s result in a p.d. that increases by a precisely determined amount as temperature changes. Bandgap sensors such as these are made as integrated circuits, but sealed in a three-wire plastic package like that of an ordinary transistor. The small package can be inserted in small spaces, has little effect on their temperature, and responds rapidly to changes in temperature.

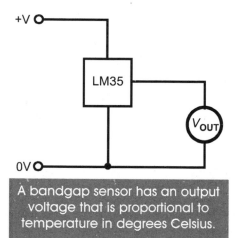

A bandgap sensor has an output voltage that is proportional to temperature in degrees Celsius.

Connected as shown in the diagram, the output of the sensor ranges from 0 V to 1.1 V as the temperature ranges from 0°C to 110°C. The response is linear, so that if, for example, we find that the output is 0.37 V, we know that the temperature is 37°C. The output is measured either by connecting a voltmeter to the circuit as in the figure, or feeding the output to a more complicated measuring and recording circuit, or indirectly to a computer.

The title photograph of this chapter shows a passive infra-red detector, used to trigger floodlights when an intruder comes into their field of view. The actual sensor is a **pyroelectric device**.

The pyroelectric device consists of a crystal of a substance such as lead-zirconate-titanate which has been heated and then cooled in a strong magnetic field. It is sensitive to changes in the amount of infra-red falling on it. Any slight change causes a p.d. to appear between opposite faces of the crystal. This p.d. can be measured and used to trigger alarms. When used in a security system, the sensor is placed behind a plastic sheet which has a raised pattern formed on it. The sheet acts as a kind of lens, focusing radiation from the surveyed area on to the sensor. But the lens has the effect of dividing the area into radiating zones. Alternate zones are visible and invisible to the sensor. When a person or any object warmer than the surroundings moves within the surveyed area it inevitably moves from a visible zone to an invisible zone, or the other way about. This sudden change in the amount of radiation received by the sensor, causes it to trigger the alarm system. This type of sensor is also used in industrial temperature measuring instruments.

Although certain aspects of infra-red sensing clearly belong in the next chapter on optoelectronics, there are some applications in the measurement of temperature that come under the heading of heat sensing. One of the prime advantages of estimating temperature from measurements of infra-red emissions is that there need be no physical contact between the subject and the measuring instrument. Industrial measurement systems include infra-red thermometers that are shaped like a pistol and aimed at the subject. For accurate aiming, a laser beam projects a spot of light on the point being monitored. This can be used at distances of several metres and for temperature ranging between −30°C and 900°C.

Closer to home, the body temperature may be measured by a similar instrument that is pointed down the outer ear channel to sample the radiation from the eardrum. It takes only 1 second to measure the temperature, which is displayed to the nearest tenth of a degree Celsius. This is a welcome replacement for the conventional mercury thermometer with its 30-second response time and the dangers of broken glass and poisonous mercury.

A **thermocouple** relies on the Seebeck effect to measure differences of temperature. It consists of two junctions between dissimilar metals or alloys, such as iron and constantan (a copper-nickel alloy), or between a nickel-chromium alloy and a nickel-aluminium alloy. The junctions can be made by taking wires of the two metals or alloys and simply twisting the ends of the wires together. It is more reliable to solder them together. The junctions are subjected to two different temperatures. One, the **cold junction**, is placed where it is subjected to a steady reference temperature (such as inside the case of the measuring instrument). The other junction, the **hot junction**, is subjected to the temperature being measured.

Under these conditions a p.d. is generated across the pair of junctions and is proportional to their temperature *difference*. This p.d. is measured by a circuit which produces a reading on a display. A thermocouple is generally used for measuring very high temperatures of several hundred or even several thousands of degrees Celsius.

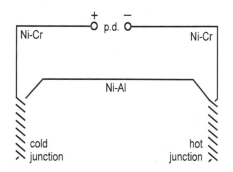

The principle of the thermocouple. A junction with the alloys shown here is known as a K-type thermocouple.

The precise temperature of the cold junction within a few tens of degrees is not significant. For high temperatures, the junctions are usually metals that can withstand high temperature without corroding. An example is a junction between platinum and a rhodium/platinum alloy.

A **thermopile** is a device in which several thermocouples are connected in series with their cold junctions grouped close together, and their hot junctions grouped likewise. This produces more current than a single thermocouple, so can be used for measuring smaller temperature differences.

Sound sensors

A useful property of crystals of quartz is that when they are slightly stressed (bent out of shape) a p.d. is developed between opposite sides of the crystal. This is known as the **piezo-electric effect** and the crystal is acting as a transducer. The p.d. produced is very small but it can be amplified to produce a usable electrical signal.

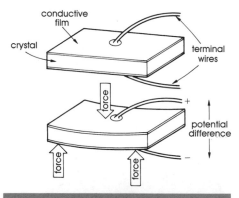

The piezo-electric effect. (Top) The transducer.(Bottom) An applied force generates a p.d. or an applied p.d. generates a force.

In a piezo-electric microphone, the sound waves impinge on a crystal or disc of piezo-electric material. They produce changes in pressure which vibrate the crystal, making it change its shape according to the shape of the sound waves. A varying p.d., corresponding to the sound waves appears at the terminals of the microphone.

Transduction by a quartz crystal operates in both directions. In the crystal microphone, the crystal is stressed and it generates a p.d. Sound (mechanical) energy is converted into electrical energy. Conversely, if a p.d. is applied to opposite faces of the crystal, it changes shape. Electrical energy is converted into mechanical energy. This effect is made use of in the small **sounders** used to produce bleeps and other sounds from digital watches and various electronic gadgets. The effect is also used in time-keeping and, incidentally, in certain types of gas-lighter. Pressure on the trigger distorts a quartz crystal and the p.d. generated is sufficient to produce an electric spark to ignite the gas.

A **capacitor microphone** has two plates separated by an air gap. One plate is a rigid metal disc and the other is a flexible metal disc or a disc of metallized plastic. Sound vibrates the thin disc altering the spacing and hence the capacitance between the two discs. The microphone requires a source of e.m.f. to charge it, but has the advantage that the sound quality is very high. A relative of this type of microphone is the commonly used **electret microphone**. This has a dielectric material between the plates. During manufacture, the dielectric is heated and then cooled while it is being subjected to a strong electric field. This causes a permanent electric field to be held in the dielectric. Vibration of the electret causes a p.d. to be generated between the two plates. Both types of capacitor microphone produce only a small signal and their output impedance (p. 185) is very high. For this reason a special FET pre-amplifier is usually included inside the microphone case.

Mechanical sensors

These are sensitive to force, to position or to motion.

A **strain gauge** consists of a thin metal foil with a zig-zag pattern of parallel conductors. The foil is mounted on a plastic backing and cemented to the object (such as a steel beam) that is under strain, with the conductors parallel to the direction of greatest stress. When a load is applied to the object, so that it becomes under strain, the conductors become stressed in the same way. They are made longer and therefore become thinner, and their resistance increases. By measuring their resistance (generally with a bridge, p. 61) we are able to determine the amount of strain.

Other forms of strain gauge rely on the piezo-electric effect (p. 118). The sensor is a strip of piezo-electric material cemented to the object under strain. When the object is strained, distortion of the material causes a p.d. to develop between its opposite faces. This is measured by a suitable electronic circuit.

Strain gauges are frequently incorporated into a device known as a **load cell**. One form of load cell is a ring of metal with four strain gauges cemented to it (a). When the cell is put under compression (b), two of the gauges are compressed and the other two are under tension. The resistances of the gauges are measured, the force calculated. The opposite occurs if the cell is placed under tension.

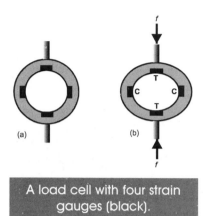

A load cell with four strain gauges (black).

Load cells are mainly in use for measuring force or torque. They are often used in automatic weighing machines and weighbridges calibrated in hundreds or thousands of kilograms, though load cells for a few kilograms are also obtainable.

Strain gauges may also be used in **pressure sensors**. A typical pressure sensor has two chambers separated by a diaphragm. One chamber is connected to a reference pressure (either gaseous or liquid) or a vacuum, or is open to the atmosphere. The other is connected to the vessel in which pressure is to be measured. The difference between pressures on the two sides of the diaphragm causes it to bulge one way or the other. The bulging is measured by strain gauges cemented to the diaphragm and the actual pressure difference in bar or other convenient units can be calculated.

In the most recent pressure sensors, the diaphragm is a wafer of silicon. Using techniques similar to those used in manufacturing integrated circuits (p. 162), a piezo-electric strain gauge is built up on the wafer, complete with bridge and amplifiers. The device may include a thermistor, with a circuit to compensate for variation in output due to temperature changes.

In **capacitive pressure sensors**, a metal plate is mounted on the diaphragm and forms one plate of the capacitor. The other plate is mounted parallel with it in a fixed position. Movements of the diaphragm cause the capacitance to vary and the variations are measured by a suitable circuit.

The simplest form of motion detector is a **mercury switch**. It is also a position or vibration detector. It consists of a metal capsule into which are sealed two contact wires and a quantity of mercury. When the capsule is upright and stationary, the mercury settles in the bottom of the capsule and makes an electrical contact between the two wires. If the capsule is tilted, moved rapidly, or vibrated, the contact is momentarily broken. The break in contact is detected by a suitable electronic circuit.

The basis of many types of position detector is the **linear variable differential transformer**, or LVDT. An alternating current is passed through the central primary coil. It generates an alternating magnetic field which is linked to the two secondary coils by the ferrite core. The secondary coils are connected in series and the direction in which they are wound is such that, when the core projects into each coil by equal amounts, the induced alternating e.m.f.s are equal and opposite. They cancel out and the total p.d. across the two coils is zero. The core is mechanically attached to the machine part whose position is to be registered. As the part moves, the core moves so that it extends more into one secondary coil than the other. The e.m.f.s are then unequal and the amplitude of the induced current increases. By measuring the amplitude of the signal it is possible to determine the position of the machine part.

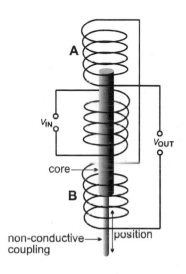

In this drawing of an LVDT, the excitation coil is coupled more strongly with coil B than with coil A.

A **linear inductive position sensor** (LIPS) is related to the LVDT but is a more recent development. The circuit is a bridge (overleaf) with two halves of the coil providing two of the arms and two capacitors providing the other two arms. A 1 MHz signal is fed into the bridge and an a.c. signal taken from the centre tap. The amplitude of the signal depends on the relative inductances of the two halves of the coil and these depend on the position of the target core. A LIPS is preferred to an LVDT because the measuring circuitry is simpler, the device is cheaper, and the range of measurable displacements is greater. A typical LIPS can measure displacements from 1 mm to 1500 mm. This contrasts with only a few millimetres for an LVDT.

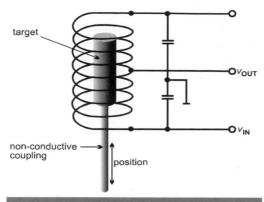

The amplitude of the output of the LIPS
depends on the position of the target.

Magnetic field sensors

If a bar of semiconducting material is placed in a magnetic field, the charge
carriers are deflected. The effect is to produce a p.d. across the semiconductor.
This is known as the **Hall effect**.

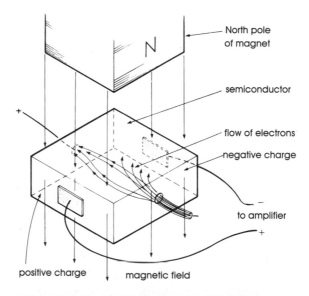

Deflection of the electrons produces a p.d.

An ampliflier is connected so as to detect the p.d.. A Hall-effect detector incorporates the semiconductor and amplifier in a three-terminal package which looks just like an ordinary transistor. Usually the sensor and amplifying circuit are integrated on the same chip. Detectors may have different types of response. Many act as on-off switches, which have the advantage of being bounce-free (p. 221). The potential at their output terminal swings abruptly from 0 V to the supply voltage when the magnetic field changes. Some of these respond to the field direction; a field in one direction turns them on, then they stay on until they are subjected to a field in the opposite direction. Others are sensitive to field strength; they are turned on when the field exceeds a certain strength. They stay on until the field strength falls below a lower level.

Hall-effect sensors are often used in **tachometers**, for measuring the rate of revolution of an axle or wheel. If the rotating part of the machine is of ferrous metal and there is a notch in it, a Hall-effect device placed close to the rotor is affected when the notch passes by. Its output is an alternating signal and the frequency of that signal is equal to the rate of rotation. A permanent magnet placed behind the device is used to increase the local field strength, making the device more sensitive to the passing notch.

Other types of Hall-effect device have a linear output. The output voltage is proportional to the field strength. This type can be used for precise position measurement of a moving part of a machine which has a pair of magnets attached to it.

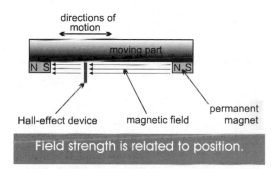

Field strength is related to position.

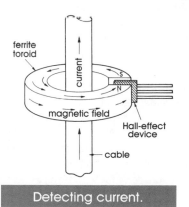

Detecting current.

Linear Hall-effect devices are used to measure large currents. A current (a.c. or d.c.) passing along the cable generates a magnetic field proportional to its strength. The magnetic field is confined to the toroid of ferrite. A Hall-effect device positioned in the gap in the toroid has an output proportional to the field strength and hence proportional to the current in the cable.

Humidity sensors

One form of humidity sensor is based on a film of hygroscopic material, which absorbs water in humid conditions and loses it in dry conditions. At any given humidity, the amount of water present in the material reaches an equilibrium. This affects the resistance between two metallic electrodes on the surface of film. The resistance is measured by a circuit which then displays the result as a reading of humidity.

In another type of sensor, the hygroscopic material forms the dielectric of a capacitor. Changes in humidity result in measurable changes in capacitance.

pH sensor

A pH meter measures the degree of acidity or alkalinity of a solution. The pH scale runs from pH0 for a very strong acid to pH14 for a very strong alkali. Water, which is neutral (neither acidic or alkaline) has a pH of 7. Two platinum-wire electrodes are immersed in the test solution. One electrode is surrounded by a standard solution containing dissolved salts. The electrode and solution are separated from the other electrode by a bulb of thin glass, which allows hydrogen ions to pass through it but restricts the movement of other ions. In effect, this is a *cell* in which the p.d. developed between the electrodes is related to the hydrogen ion concentration in the solution. This is directly related to the pH. A circuit measures the p.d. between the electrodes and displays the pH on a scale.

Gas sensors

Gas sensors are used for detecting a wide range of gaseous substances in the atmosphere, including pollutants, toxins and combustible gases. In the case of combustible gases, the sensing element is a platinum wire coated with a mixture of substances which catalyse the oxidation of the gases. The coil is heated by passing a current through it; this causes combustion to occur, generating additional heat and thus increasing the resistance of the coil. The circuit also includes a dummy element from which the catalysts are absent. This too is heated by passing a current through it, but does not gain the additional heat from gas combustion, so its resistance is not raised. A bridge circuit compares the resistances of the two coils. When gas is present, the bridge goes out of balance and this may be used to trigger an alarm. In practice the two coils are contained within a dome of stainless steel mesh. This is to prevent combustion around the sensing element from spreading to the atmosphere outside, possibly causing an explosion.

Combustible gases that are detectable with gas sensors include methane (present in town gas and natural gas), butane, propane, hydrogen, and alcohol vapour. Sensors have also been developed for ozone, nitrogen dioxide, chlorine, ammonia, and toluene. Sensitivity varies with the gas, but a typical sensor can give a reading in the range 0 to 5 parts per million.

Nuclear radiation detectors

A wide range of radiation detectors has been invented, including the ionization chamber, the scintillation counter and the Geiger-Muller tube, but we describe a more recent solid-state device which makes use of the properties of semiconductors. Like the Geiger-Muller tube, it depends for its action in producing an avalanche of charge carriers, which generates a pulse. The **extrinsic semiconductor nuclear radiation sensor** consists of a slice of n-type silicon, the surface layer of which is doped to convert it to p-type. The upper and lower surfaces of the slice are metallized to form electrodes. In operation, an external circuit produces a p.d. across the device.

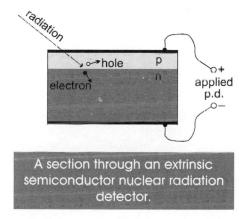

A section through an extrinsic semiconductor nuclear radiation detector.

Nuclear radiation passing into the p-type layer displaces electrons from the atoms, creating electron-hole pairs. These are accelerated by the electric field within the slice, creating further electron-hole pairs. The result of the arrival of nuclear radiation is thus a sudden increase in the number of charge carriers and therefore a pulse of current through the device.

The pulse is detected by the external circuit. The rate at which pulses occur is proportional to the amount of radiation being received.

Sound-producing transducers

Most loudspeakers work by electromagnetism. A powerful magnetic field is produced by a specially-shaped permanent magnet, rigidly attached to the chassis of the speaker. A lot of research has gone into developing materials that can provide a high-intensity field, and are not readily demagnetized.

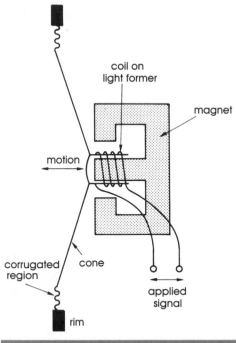

coil on
light former

magnet

motion

corrugated
region

cone

applied
signal

rim

A section through an
electromagnetic loudspeaker.

The voice coil is a coil of thin wire usually wound on a light plastic core. It is attached to the rear of a paper or plastic (mylar) cone. The code and coil are supported by a ring of corrugated paper or plastic, which allows the cone to vibrate backward and forward with great freedom. When a signal current from an amplifier is passed through the coil, a varying magnetic field is generated within the coil. The interaction between the field of the permanent magnet and that of the coil, forces the coil to vibrate. These vibrations are transferred to the cone which, in turn, makes the air in front of the cone vibrate, generating sound waves in the air.

A loudspeaker is an inductive device so its impedance depends on frequency (p. 52). The impedance of a loudspeaker coil is generally quoted for a frequency of 1 kHz. Most loudspeakers have low impedance, 4 Ω or 8 Ω being common values. The low impedance allows a large current to flow, so that the speaker has a high power rating (p. 33) and the sound produced is loud. Small speakers of higher impedance (often 64 Ω) are sometimes used in battery-powered radio sets and tape players.

The other common way of converting electrical energy into sound is by a piezo-electric **sounder**. Electronic sirens and buzzers have a sounder of this type, with built-in transistor circuits to generate the alternating p.d. required.

Motion-producing transducers

Probably the most familiar example of a device which converts electrical energy into motion is the electric motor.

The operation of electric motors is more a concern of electrical engineering rather than of electronics, so we will not describe these here. However, there are two specialist types of motor that are particularly suited to be controlled by electronic circuits so these deserve a mention. A **stepper motor** has four sets of coils arranged so that the rotor is turned from one position to the next when the coils are energized in fixed sequence. The energizing of the coils is easily achieved by a digital electronic circuit. Typically the rotor turns 15° at each step, but there are motors that turn only 7.5° or even a smaller angle. The motor can be single-stepped in either direction or it can be stepped continuously at a rate that determines its rate of rotation. Yet it can instantly be brought to a halt in any one of its 24 positions. Stepper motors are of great use in industrial machinery, including robots.

A **switched reluctance motor** has more poles on its stator than on its rotor. Only one pair of poles can be aligned at any one time. The rotor, which does not have coils on it, is made to turn at a controlled speed by switching pulses into the coils of the stator.

The pulses need to be specially shaped, which is done by employing a special controller chip, or by a microcontroller circuit. The motor can be run in either direction at a precisely controlled speed and can be put through a sequence of controlled acceleration and deceleration. This makes it very suitable for driving industrial machinery including robots and also, for example, moving the flight surfaces of aeroplanes.

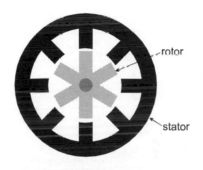

rotor

stator

The stator (the stationary part) has eight poles but the rotor has only six.

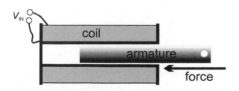

A solenoid, seen here in section, exerts a strong force when a current is passed through its coil.

The simplest motion producer is a **solenoid**, which consists of a coil with a soft iron core (or armature) which slips easily in and out of the coil. Soft iron is used because it does not retain its magnetism when the current is switched off; in other words, it does not become permanently magnetized.

To start with, the armature lies with only a part of it inside the coil. When a current is passed through the coil, the armature is drawn fully into the coil with great force. This gives solenoids many uses in driving parts of robot machines, or operating bolts on doors and windows. Turning off the current does *not* force the armature out again, so the reverse action must be provided for by some external mechanism. This may be simply a spring, compressed when the solenoid is energized and released when the current is cut off. In robots, the reverse action may be effected by a second solenoid operating in the reverse direction.

A **relay** is an electromagnetically operated switch. It consists of a coil wound on a soft-iron core, and a soft-iron armature mounted on a pivoted lever. The lever has contacts on it which close against other stationary contacts mounted on the chassis of the relay. The arrangement of contacts depends upon the model of the relay. Usually there is a pair of changeover contacts as shown in the figure.

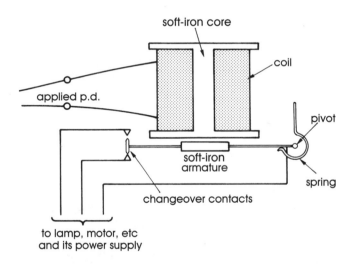

When a current passes through the coil, one contact is made and the other one is opened.

When there is no p.d. across the coil, the movable contact closes against the lower fixed contact. This pair of contacts is described as being **normally closed**. There is an open circuit between the movable contact and the upper fixed contact. It is described as **normally open**. The situation is reversed when a p.d. is applied to the coil. The movable contact is moved away from the lower fixed contact and closes against the upper contact.

Relays are normally used for controlling lamps, motors and similar devices. A small current, perhaps that provided by a transistor switch or the output from a logic gate, may be used to control the very much larger current switched through the relay. The changeover contacts make it possible to turn one device on when another is turned off. Some relays have a more complicated array of contacts, allowing several devices to be switched on or off simultaneously. Although junction transistors and MOSFETs can also be used for switching, a relay has the advantage that it can have contacts suitable for operating at high voltages, with high currents. A particular advantage is that relays can switch alternating currents (such as the mains), which is not possible with transistors.

Relays are made in a number of sizes, including miniature types suitable for mounting on a printed circuit board (p. 159). Some of these are contained in a package identical to that used for integrated circuits.

A **reed switch** has the same function as a relay. Reed switches are not able to carry such heavy currents as a typical relay, but they act more quickly. The switch consists of two springy contacts sealed in a capsule to exclude dust and dampness. The ends of the contacts overlap but do not touch. In one type of reed switch, there is a coil wound round the capsule. When a current is passed through the coil the contacts become magnetized temporarily. They are both magnetized with the same polarity so that the north end of one contact is close to the south end of the other. They are attracted toward each other and meet, completing an electric circuit. When the current to the coil is turned off, they lose their magnetism and spring apart.

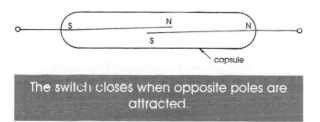

The switch closes when opposite poles are attracted.

A reed switch (without a coil) may also be operated by bringing a permanent magnet close to it. This is useful for monitoring the action of machinery; the magnet is attached to a moving part and one or more reed switches are fixed where they can detect the position of the magnet. Reed switches are used in security systems to detect if doors and windows have been opened. To protect a door, a reed switch is mounted in or on the door frame. A permanent magnet is mounted in or on the door itself so that, when the door is shut, the magnet comes close to the switch and the switch closes. When the door is opened, even by as little as a centimetre, the contacts spring open.

A number of such switches on the doors and windows of a building may be connected in series to form a loop. If all doors and windows are shut, all switches are closed and current passes round the loop. But if any one door or window is opened, its switch contacts open too, breaking the loop. This breakage is detected by a centrally-located circuit, which immediately triggers an alarm. The alarm is sounded also when the loop is broken by an intruder cutting the wires.

12 Optoelectronic sensors

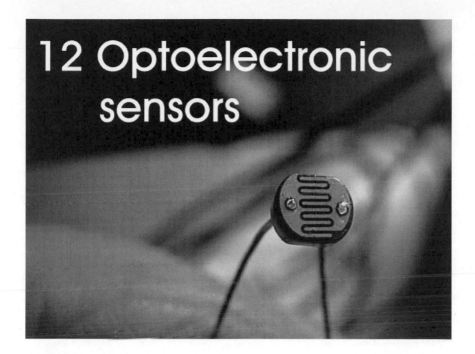

There are so many devices that are sensitive to light, and so many applications for them, that we are devoting a separate chapter to the optoelectronic sensors.

Light-dependent resistor

This consists of a disc or square block of semiconducting material, usually cadmium sulphide, with two electrodes printed on its surface (photograph above). The electrodes are shaped so that there is a narrow gap between them, but their edges are waved to make the length of the gap as great as possible. The whole is encapsulated in clear plastic. When a p.d. is applied to the electrodes, a current passes from one electrode to the other.

Under dark conditions, the current is limited by the small quantity of charge carriers present in the semiconductor. The resistance of the device is high, perhaps as great as 10 MΩ. When light falls on the disc, its energy excites electrons in the atoms of the semiconductor and many electron-hole pairs are generated. This greatly increases the number of charge carriers and the current increases. Putting it another way, the resistance decreases. In bright light it may fall as low as 100 Ω.

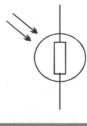

Symbol for an LDR.

Because the resistance of the device depends on the intensity of light falling on it, it is known as a **light-dependent resistor**, often shortened to LDR. Another name for this device is **photo-conductive cell** but, unlike other cells, it does not generate a p.d., so this name is misleading.

LDRs have many applications for measuring light levels, for example, in the exposure controlling circuit of a camera. They operate in a wide range of light intensities. Their main disadvantage is that their response time is low. An LDR takes about 75 milliseconds to respond to a fall of light intensity (the time required to sweep the existing charge carriers away into the external circuit) and they may take several hundred milliseconds to respond to a rise in intensity. This is no disadvantage in some applications but in others, such as reading a bar-code (p. 298), their response is too slow.

Photodiode

The p-n junction of an ordinary diode is sensitive to light. Usually, light is excluded by sealing the diode in a light-proof package of plastic or painted glass. A photodiode has a transparent cover; it may either be a glass capsule or a metal can with a lens at its end to focus light on to the p-n junction. Infra-red photodiodes have a package that is opaque to visible light but transparent to infra-red.

The effect of light is to excite the electrons in the atoms of the semiconductor, causing some of them to escape, leaving holes. The increase in electrons and holes means that there are more charge carriers and, for a given applied p.d., the current increases. In most circuits the photodiode is reverse biased, so that charge is carried by extrinsic charge carriers. The response time of a photodiode is typically 250 nanoseconds. Compare this with the response time of an LDR, measured in milliseconds, a unit a million times longer. Photodiodes, particularly the infra-red variety, are used in security systems and in TV remote control systems.

PIN photodiodes have a layer of intrinsic (not doped) semiconductor between the p-type layer and the n-type layer. The effect of the intrinsic layer is to reduce the capacitance between the p-type and n-type layers (p. 40 and p. 79). This makes it easier for the diode to respond more rapidly to changes in light levels. Response times of the order of 0.5 nanoseconds are typical. This is important in diodes being used to receive high-frequency signals by optical fibre. Avalanche photodiodes are sensitive to very low light levels and, like PIN photodiodes have response times in the region of 0.5 ns.

Phototransistor

This has a similar structure to an ordinary npn junction transistor, except that the case has a glass window, usually shaped like a lens to focus light on to the transistor. The cheaper types are encapsulated in transparent plastic. As the symbol indicates, a phototransistor does not necessarily have a connection to its base layer. When light falls on the transistor, atoms in the collector region become excited and emit free electrons. This creates electron-hole pairs. These electrons leak through the reverse-biased collector-base junction. Once in the base region they pass on to the emitter; they are the equivalent of a base current. This base current is amplified by the usual transistor action (p. 88), resulting in an increase in the collector current. Some types of phototransistor have a connection to the base layer. This makes it possible to bias the transistor (p. 184) so that its response time is reduced.

The amplification due to transistor action means that phototransistors are more sensitive than photodiodes. But their response time is much slower, usually several milliseconds. With the development of integrated circuit techniques, it is now possible to manufacture a photodiode and an amplifier circuit on the same chip. This combines high sensitivity with a short response time. **Optoschmitt detectors** combine on the same small chip a phototransistor, amplifier, voltage regulator, Schmitt trigger (p. 101) and an output stage suitable for driving logic circuits.

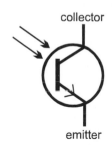

collector

emitter

Symbol for a phototransistor that has no base terminal.

Looking through the lens at the top of the phototransistor case, we can see the silicon chip with the transistor fabricated on it.

Photovoltaic cell

Unlike the other devices described so far in this chapter a photovoltaic cell generates a p.d. when light falls on it. This makes it a transducer as well as a sensor (p. 114). The early photovoltaic cells consisted of an iron plate coated on one side with selenium (a semiconductor), and were known as **selenium cells**. They were used in early motion picture projectors, to read the optical sound.

Nowadays, photovoltaic cells are based on silicon as the semiconductor. A typical p-on-n photovoltaic cell has a very thin layer of p-type silicon diffused into one surface of a block of n-type silicon. The p-type layer is so thin that light penetrates it and reaches the depletion region. The virtual cell at the p-n junction produces a p.d. of about 0.6 V across this junction. The n-type layer is positive with respect to the p-type layer.

When light falls on the cell, the atoms are excited and liberate electrons, creating electron-hole pairs. The electrons are attracted toward the positive n-type layer by the field of the virtual cell. Conversely, the holes are attracted toward the p-type layer. If the two layers are connected through an external circuit, the extra electrons in the n-type later pass through the circuit to combine with the extra holes in the p-type layer. In other words there is a current in the external circuit. This continues to flow for as long as light is falling on the cell and renewing the supply of electrons and holes. The p.d. across the cell is about 0.6 V, due to the virtual cell but, now that there is a source of energy (light), the virtual cell is able to supply a continuous current. The current is proportional to the intensity of the light falling on the cell.

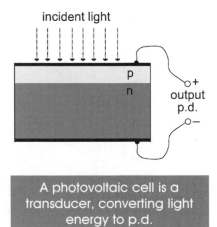

A photovoltaic cell is a transducer, converting light energy to p.d.

Photovoltaic cells are made with a relatively large area to produce a reasonably large current. Response time is short (typically 50 ns) and response is linear. This makes them useful in optical instruments and also for optical-fibre communications. One type is made that incorporates a colour filter in its lens system to give it a similar response to that of the human eye. This has applications in colour-matching instruments.

Charge coupled device

Charge coupled devices, or **CCDs**, have applications for storing data in computer systems, but their more significant application is the detection of optical images. They are widely used in TV and video cameras, in scanners, and have made possible the introduction of digital cameras that promise to eventually replace the traditional film cameras.

A CCD consists of a slice of p-type silicon with a thin layer of n-type on one surface. This layer is coated with an insulating layer of silicon dioxide, on top of which is deposited an array of metal gates. The structure is similar to that shown on the opposite page except that the polarities of the layers are reversed and that the upper metallized layer has a complex pattern instead of being a continuous layer. Electric charge can be injected into the n-type layer (in memory applications) or generated there when light falls on the chip and produces electron-hole pairs (in video applications). If a gate electrode has a positive potential applied to it, the charge can be confined beneath the gate in the n-type layer. The unit acts as a capacitor, storing a variable amount of charge, which can be retained for several hours. The amount of charge stored depends on the amount of light falling on that region of the chip.

A digital camera (like a computer monitor) works by dividing an image into numerous (perhaps two or three million) **picture elements**, usually known as **pixels.**

A CCD has an array of rows and columns of pixel units, each consisting of a number (usually three) electrodes. An image of the scene is focused on the CCD chip and the units accumulate charge in proportion to the light intensity. In a colour system, there are ways of sampling the three colour regions, red, green and blue separately but we shall confine our discussion to shades of grey.

The array of pixel units may be represented by the diagram overleaf. Each unit has a central electrode which confines the charge to the area of the pixel when its potential is positive. The charge may be transferred to an adjacent unit by a sequence of changes in potential of this central electrode and the other electrodes in the unit. The charge is transferred with only 0.001% loss.

Individual pixels from an area of the photograph on page 133.

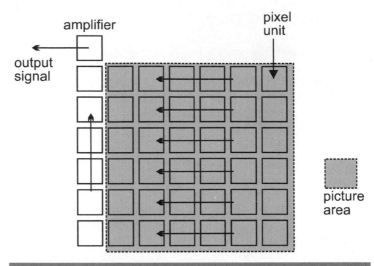

In a CCD , the charge in each pixel unit is passed along the rows and, at each step, the charges of the end units are passed to the output amplifier, to be read out.

The pixel units sample the light intensity at a million or more points in the picture area, arranged in rows and columns. The pixels in the rows are linked to their neigbours so that, when the time comes for reading out the image, the charge in each unit is transferred to its neighbour. The electrodes are connected so that the charges are shifted one step along the row at each step. At each step, the charges in the end units (at the left of each row in the drawing above) are shifted into a row of units running perpendicular to the other rows. A sequence of shifts in this row feeds the contents of the cells in order to an amplifier, where the charges are measured and an output signal is generated. The output may be an analogue signal, or in digital cameras is converted into a digital signal. This may be a 24-bit signal which covers over 16 million different tones in all colours. After the charges in the end units have been shifted to the amplifier and measured, the charges in the rows are shifted one more step to the left and the process is repeated. The output from the CCD is transferred to the memory of the camera where it may be used to produce the viewfinder image. It may later be downloaded into the memory of a computer, which may display the image on the screen, send it to a colour printer, or attach it to an e-mail and send it to another computer.

CCDs are highly sensitive to light, the most sensitive types being able to detect as few as 10 photons. In terms of photographic sensitivity, the cheaper CCDs have sensitivity in the range 150ISO to 400ISO, while the more expensive ones rate up to 1600ISO.

Solar cell

This has the same essential structure as a photovoltaic cell and works in the same way. The chief difference is that a solar cell has a very much larger area than an ordinary photovoltaic cell. Instead of a continuous n-type layer the electrode consists of a pattern of strips. This arrangement allows more light to penetrate to the depletion layer, and offers less resistance to the flow of current to the terminal. Solar cells are usually connected in series to produce a battery with high total p.d. Batteries with an output p.d. up to 30 V can be purchased ready-made. Under full sunlight, these deliver a current of several hundred milliamps. Solar batteries are used to provide power for equipment that is remote from mains power supplies. Microwave link transmitters on mountain tops are a good example. In countries where sunshine is plentiful, solar cells are used to power emergency radio-telephones along freeways. They are also used for powering spacecraft and satellites. Experimental vehicles have been successfully powered by solar batteries, including an aeroplane which obtains power from solar batteries located on its wings. Moon buggies are another example of vehicles powered in this way.

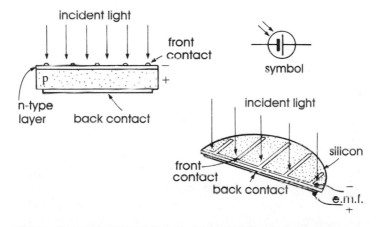

Photovoltaic cells are often made on half of a circular slice of silicon.

Unfortunately the efficiency of solar cells is low. They convert only 15% of the light energy falling on them. Research is being succesful in improving efficiency and reducing cost.

13 Light sources and displays

One of the most familiar of light-producing transducers is the filament lamp. Current passing through the filament causes it to glow white-hot. Some of the electrical energy is converted to heat, but much is converted to light. Although electronic circuits may sometimes include filament lamps as indicators, the light source most representative of electronics is the light-emitting diode.

Light-emitting diode

The most typical form of a light-emitting diode, or LED is illustrated above. The case is made from transparent plastic, usually in the same colour as the LED, though clear plastic is used sometimes. The most common colour is red, since red LEDs produce more light than LEDs of other colours. Other colours include orange, yellow, green and blue. More recently, white LEDs have been produced, consisting of a blue LED which contains a phosphor that glows white when stimulated by the blue light from the LED.

Energy is released when the electrons in the n-type material of a diode meet the holes in the p-type material. In a silicon or germanium diode this energy is lost as heat. If the material is doped with compounds containing gallium arsenide, the energy is released as visible light instead. The colour of the emission depends upon the exact composition of the dopant.

Compounds based on gallium arsenide have electrons of relatively low energy and so emit radiation of the longer wavelengths, particularly infra-red and red, but also yellow and green. LEDs containing gallium nitride are also made for higher luminance. Blue LEDs are based on silicon carbide or, for higher intensity, indium gallium nitride.

Infra-red LEDs are used to produce invisible beams in security systems; the beam is directed at a sensor, usually an infra-red photodiode. When the beam is broken by an intruder passing between the source and the sensor, the alarm is sounded. Infra-red LEDs are also used in TV and video player remote control systems. The infra-red LEDs in the hand-held controller are used to flash a coded series of pulses. The pulses are detected by an infra-red photodiode in the TV set and decoded. This causes a change of programme, increase of volume or other action depending on which button was pressed on the controller.

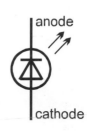

anode

cathode

LED symbol.

LEDs are made in a range of sizes from the miniature ones only 2 mm in diameter to the jumbo-sized LEDs which are 10 mm in diameter or more. They are also made in a variety of shapes: circular, triangular, rectangular. A row or 'stack' of LEDs placed close together is used to make a bar graph. These are often found on audio equipment to indicate sound volume. At low volume, only the bottom few LEDs of the bar graph are lit but, as volume increases, more and more of the LEDs light up.

LEDs are used to display letters of the alphabet or numerals. One way of doing this is to arrange LEDs in a rectangular array of, say, seven rows of five LEDs. Different characters are produced by turning on the appropriate LEDs. A row of such arrays is used for displaying messages.

These 'seven-segment' displays actually have eight segments, the central one being divided into two shorter segments.

If only numerals and a few letters are required, the **seven-segment display** is the simplest and most popular. Seven rectangular LEDs, and usually one or two smaller square LEDs, are mounted in a plastic block. By turning on the appropriate segments any numeral can be displayed. The small squares are used as decimal points or to form a colon to separate hours from minutes.

Optocoupler

Optocouplers are an important application of LEDs. An LED and a phototransistor are sealed in a light-proof plastic package, so that light from the LED is received by the phototransistor. When the LED is turned on by a current supplied from an external source, the phototransistor is turned on. If the phototransistor is wired as a switch, this can turn on other devices.

The device acts as a relay but operates considerably faster. Such devices are used when signals must be sent from one circuit to another but it is not suitable for the circuits to be electrically connected. For example, the LED may be switched on and off by a logic circuit operating at logic voltages (for instance 3 V or 5 V), while the transistor may be part of another circuit in which it is operating at, say, 12 V. Or it may be essential for two circuits to have their 0 V rails to be isolated to prevent interference being carried across.

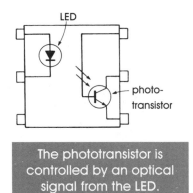

The phototransistor is controlled by an optical signal from the LED.

In a similar device, known as an **opto*isolator***, the LED and photodiode are electrically insulated from each other to withstand voltage differences of several thousands of volts. Optoisolators are generally used for *safety* reasons. For example, to protect an industrial microcomputer from the dangers of large voltage spikes generated in industrial electrical switch-gear. Another application is where one part of a circuit is to be connected to the human body, and the other part of the circuit operates at dangerously high voltage.

As well as simple 'on-off' signals, the optocoupler can transmit continuously varying analogue signals. Optocoupled thyristors and triacs are used to switch alternating currents at mains voltages by means of low-voltage control circuits.

Laser

A laser is a device in which a large number of atoms are excited in such a way that they all emit radiation of a single wavelength in phase with each other. A laser produces a beam of light of very high intensity and with high energy content. Lasers are used to produce high concentrations of energy in a small area, in order to generate extremely high temperatures and pressures.

Because laser light is monochromatic (a single wavelength) it can be focused into a narrow pencil of light. Laser beams may travel for many kilometres without appreciable widening. Laser beams have been projected to the Moon and reflected back to Earth. The laser is not an electronic device so a description of how it works is outside the scope of this book.

Laser diode

A laser diode is similar to an LED in many respects but it produces a narrow beam of high intensity. The radiation produced by a laser diode (usually in the red or infra-red regions of the spectrum) is of a single wavelength so that, if it is passed through glass panels or lenses, the beam is refracted as a whole, not split into different wavelengths. When the light is produced by the diode element it spreads out over a wide angle, but is easily focused to form a narrow and virtually parallel-sided beam. The beam of a typical laser diode is 4 mm × 0.6 mm, widening only to 120 mm at a distance of 15 m. The ability to produce a very narrow beam makes the laser diode specially suitable for use in bar code readers (p. 298) and in CD and CD-ROM players (p. 234). Since a laser diode can be switched on and off at high frequencies, even as high as 1 GHz, they are also used for telecommunications by optical fibre (p. 268).

Liquid crystal display

A liquid crystal display consists of two glass plates with a thin layer of liquid crystal material sandwiched between them (overleaf). The inner surfaces of the glass are printed with a transparent film of conductive material, such as indium oxide or tin oxide, to form electrodes. The liquid crystal material is an organic substance that is liquid but, unlike most liquids, has crystalline properties. Its molecules are long and narrow in shape and the naturally-occurring forces between them tend to make them lie together in a regular pattern. In an LCD they normally lie parallel with the glass plates. This gives the material its crystal-like properties.

The most important effect of its crystalline nature is that the material rotates the plane of polarization of light as it passes through the crystal. As the figure shows, the material is of the thickness required to rotate the plane of polarization by exactly a quarter of a turn. If films of polarizing material are placed above and below the display, with their polarizing directions at right-angles, light entering at the top is polarized, its plane is rotated 90°. It is then able to emerge through the bottom polarizer. Light apparently passes unaffected through the display.

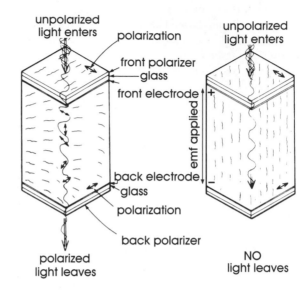

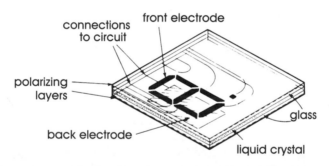

When no field is applied, light passes freely through the layers (left). When a field is applied between the front and back electrodes the light is prevented from passing through the layers (right). The lower drawing shows how shaped front electrodes define the display.

If a p.d. is applied between the electrodes, the electric field alters the arrangement of the molecules. They lie perpendicular to the electrodes. As a result, there is no rotation of the plane of polarization in the crystal. Light passing in through the top polarizer remains polarized in its original direction and is unable to pass out through the bottom polarizer. The liquid between the electrodes appears to be black. If the p.d. is removed the molecules immediately return to their crystalline arrangement and the display becomes transparent again.

In a liquid crystal display (LCD) the back electrode (or back plane) usually covers the whole inner surface. A pattern of separate electrodes is printed on the front inner surface. This takes the form of one or more seven-segment arrays and a variety of other symbols, such as maths signs and legends, depending on the application. Very narrow tracks pass from each segment to the edge of the display and are there connected to a terminal strip. These tracks are too narrow to show up as black lines. When a p.d. is applied between any one or more of these segments and the back plane, the area appears black. By switching on different combinations of segments, we can generate numerals or alphabetic characters.

An indoor/outdoor thermometer uses an LCD to display the temperature at two sites. It runs for months, day and night, on a single AAA size dry cell.

The illumination for an LCD comes from external sources such as daylight or a built-in filament lamp. The display may be viewed by reflected light, with a mirror behind it to increase contrast, or it may be viewed by transmitted light, with an illuminated panel behind it. The LCD display has the advantage that it can be viewed in full daylight, whereas an LED display is not bright enough. On the other hand, an LED display has the advantage in darkness. Since it produces its own light, an LED display requires an appreciable current, usually several tens of milliamps. By contrast, the current required by an LCD is only a few microamps. This is an advantage for displays on calculators and other portable battery-powered equipment (see photograph), particularly for lap-top computers.

Cathode ray tubes

These are widely used as displays in oscilloscopes (p. 150), radar equipment (p. 293), TV sets and computer monitors. The way they work is described on page 150.

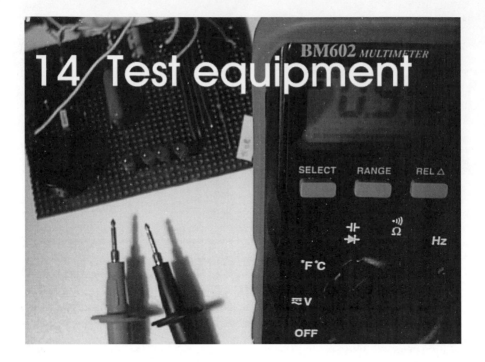

14 Test equipment

The term *test equipment* covers everything from a simple gadget for checking continuity to a computer-controlled rig that automatically checks the operation of a newly manufactured DVD player. Here we look at the principles of circuit testing, describing the equipment that might be used by a typical service engineer.

The moving-coil meter

This moving-coil panel meter is used to measure and display a voltage.

Nowadays, most meters have a digital readout, but the old-fashioned moving coil meter, with its needle moving over a graduated scale, still has many uses and is preferred by some people. Its operation relies on the fact that if a coil carrying a current is placed in a magnetic field, a force will act upon the coil. The coil (see opposite) is wound on a former of aluminium over a cylindrical iron core. The former is mounted on pivots so that it can rotate between the poles of a permanent magnet.

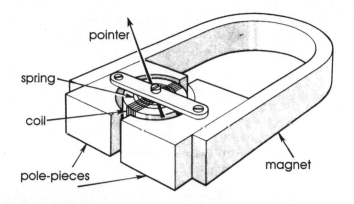

pointer

spring

coil

pole-pieces

magnet

The movement of a moving-coil meter.

There is only a narrow air gap between the poles and the former, so the magnetic field in which the coil turns is radial in direction and uniform in strength for any position of the coil. The pointer is carried by the coil. The turning motion of the coil is opposed by a phosphor bronze spring.

When a current passes through the coil, the coil tends to turn on its axis against the opposing force of the spring until it takes up an equilibrium position which is proportional to the current flowing through the coil. As the coil turns, the pointer attached to it moves across a graduated scale. The amount of turn is proportional to the current passing through the coil so the scale is graduated by comparison with a standard current meter to show the current, usually in microamps or milliamps.

Measurement of current

The elementary moving-coil ammeter that we have just described may have its current range extended by using a **shunt**. This is a low-value resistor wired in parallel with the coil of the meter. To see how it works, suppose that we have a meter in which the needle is deflected the whole way across the scale when a current of 1 mA is flowing through the soil. We say that the **full scale deflection** (or f.s.d.) of the meter is 1 mA. In the figure, such a meter is shown with a shunt across it. The resistance of the shunt is 1/1000 of the resistance of the coil.

When a current reaches the junction between the meter and the shunt, 1/1000 of the current passes through the coil and 999/1000 of it passes through the shunt. If the needle is fully deflected, we know that 1 mA is passing through the coil. But this is only 1/1000 of the current in the circuit. So the current in the circuit must be 1000 times greater, which is 1 A. The shunt has converted the 1 mA meter to a 1A meter.

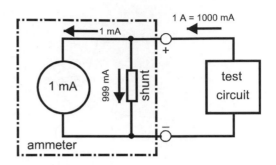

An ammeter with f.s.d. = 1 A consists of a milliammeter wired in parallel with a shunt resistor. The currents shown are the maximum currents for this meter.

Note that a current meter (an **ammeter**) has very low resistance, equal to that of the coil and shunt in parallel. When we are measuring current we connect the meter, and its shunt if there is one, so that all the current flowing in the circuit must pass either through the coil or through the shunt.

Measurement of p.d.

Given a meter coil with resistance R and given that the meter (in this example) has an f.s.d. of 50 µA, we can calculate from Ohm's Law (p. 25) that, when the needle is at full-scale, the p.d. across the meter is:

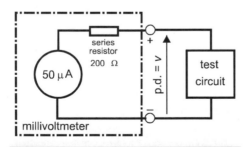

A voltmeter with f.s.d. = 10 mV consists of a milliammeter wired in series with a 200 Ω shunt resistor.

$$V = 50 \times 10^{-6} \times R$$

If the coil resistance is 200 Ω, the f.s.d. is $v = 50 \times 10^{-6} \times 200$ = 10 mV. We have a meter capable of measuring p.d.s up to 10 mV. This is a very limited range but we can easily extend it by putting a higher valued resistor in series with the meter coil.

Contrast this series resistor with a shunt resistor, which is a *low*-value resistor in *parallel* with the coil. Suppose that the value of the series resistor is $R = 99$ times that of the coil. Given that the same current passes through the coil and through the resistor, the fall of p.d. across the resistor is 99 times that across the coil. The fall in p.d. across the coil and resistor together is 1000 times that across the coil alone. At f.s.d., when the p.d. across the coil is 10 mV, the p.d. across coil and series resistor is $1000 \times 10 = 10\,000$ mV = 10 volts. We have increased the range of the meter to a f.s.d. of 10 V. By using series resistors of other values we can obtain any f.s.d. we require.

Note that a p.d. meter (a **voltmeter**) has very high resistance, equal to that of the coil and series resistor in series. When we are measuring p.d., we connect the meter, and its series resistor if there is one, to the two relevant points in the circuit. Only a very small current passes through the meter.

The descriptions of current and p.d. measurement apply to direct current and d.c. voltages. To convert d.c. into a.c., we wire a diode in series with the meter. Usually the frequency of the a.c. is too high for the needle to be able to follow the rapid changes in current or p.d. It shows an 'average' reading. This is less than the peak value (amplitude) and it can be shown that it is actually 0.707 times the peak. This is often known as the **root mean square** (or r.m.s.) value. Meters intended for measuring a.c. may have the scale calibrated so as to read peak and r.m.s. values.

Measurement of resistance

The simplest technique for measuring resistance has the test resistance wired in series with a battery and a milliammeter. The cell provides a known p.d., *V*. The meter measures the current, *I*. From these two values we may calculate the resistance *R*, using the Ohm's Law equation, $R = V/I$.

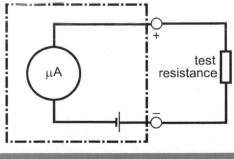

A simple way to measure resistance.

Rather than use a cell directly, as in the figure, it is better in practice if we use a Zener diode or bandgap reference to provide a stabilized p.d. (p. 108). If the value of *V* is fixed in this way, the equation shows that R is proportional to $1/I$. We can calibrate the scale of the meter directly in ohms.

Unfortunately, the relationship between resistance and current is inverse. As a result, the scale reads from 'infinite' resistance at the zero end (very large resistance means very low current) to some minimum value of resistance at f.s.d. The ohms scale is reversed and is not linear, making accurate reading difficult.

There is a source of error in this technique, for we are ignoring the resistance of the meter itself. The value R is really the sum of the resistance of the meter and that of the resistance under test. Ammeters usually have a relatively low resistance so the meter resistance can usually be ignored, but there will be serious errors if the resistor under test also has a low value.

Multimeters

Moving coil meters used in scientific laboratories are often made with a single shunt or series resistor for measurements over a single range. In electronics, we more often use a meter which has switchable ranges. This is known as a **multimeter.** The multimeter has two wires plugged into its positive and negative (0) terminal sockets. These may end in a test prod, a crocodile clip or some other type of connector. Usually there is a rotary switch, used to bring various shunts or series resistors into the measuring circuit. The meter may also have a separate input terminal for a.c. measurements with a diode leading to the measuring circuit, or there may be a d.c./a.c. switch to bring the diode into action when needed.

Digital meters

A typical digital meter is illustrated in the photograph at the beginning of this chapter. Instead of a dial with a pointer, a digital meter has a display consisting of four (in some cases more) digits, together with polarity symbols (+ and –) and a decimal point that automatically appears between the correct pair of digits. The display also shows the units of measurement, such as 'V', 'mV' or 'Ω '. The display is a liquid crystal display and is updated at intervals of a few seconds. Although the display may be the most obvious difference between this type of meter and the moving-coil meter, there is a much more fundamental difference between the two types. The input terminals of the moving-coil meter connect the test circuit to the coil and possibly a shunt or series resistor. The input terminals of a digital meter connect the test circuit to an operational amplifier (p. 192), with FET inputs. The result of this is that the digital meter draws virtually no current from the test circuit.

The actual resistance of a moving-coil meter depends to a large extent on the quality of the meter movement. A low-cost meter usually has a coil of relatively low resistance and the series resistor has a correspondingly low value. The resistance of an inexpensive moving-coil meter is 20 kΩ when switched to its 10 V range. Such a meter draws an appreciable current from a high-resistance test circuit. On page 61, we showed how connecting a low-resistance circuit to a potential divider results in a fall in potential at the connection point. The same happens when a low-cost meter is used for measuring potential in a test circuit. The meter draws excessive current from the circuit, leading to a fall in potential at the point to which it is connected. It reduces the potential that it is trying to measure, giving a falsely low reading. This effect is much less important when using an expensive meter with a high-resistance coil (typically, 200 kΩ on the 10 V range), but with *any* moving-coil meter this effect can lead to errors. By contrast, a digital multimeter with its FET input presents an input resistance of at least 2 MΩ and, more usually, as high as a million megohms. Such a meter has almost no effect on the test circuit and an accurate reading of potential is obtained.

Digital meters are the product of the latest developments in electronic technology so it is to be expected that they will incorporate many features that are not available on the typical moving-coil meter. As well as a wide selection of voltage, current and resistance ranges (including accurate low-resistance range) many of these meters also provide for measurements of capacitance and frequency. A continuity tester, which produces a 'beep' when there is an electrical connection between the probes is almost a standard feature. The meter may have a temperature probe (a thermocouple) and display the temperature in degrees Celsius or Fahrenheit. The more expensive digital meters include circuits for testing diodes and transistors, including measurement of transistor gain.

As well as indicating instantaneous readings, the more expensive digital meters have the ability to process a series of readings taken over a given period of time. At the end of the period they can display the minimum, the maximum, the difference between the minimum and maximum, and the average of the readings it has taken since it was last reset.

In spite of the advantages of the digital meter, some engineers prefer the moving-coil meter for investigating fluctuating p.d.s and currents. Useful information may be gained by watching the way the needle moves across the scale. It is far from easy to extract this information from a set of rapidly changing digits. However, the more expensive digital meters have a way of presenting this information as a bargraph. Watching the end of the bar moving to and fro is equivalent to watching the needle of the moving-coil meter.

The oscilloscope

One of the most useful test instruments in electronics is the oscilloscope, which is used to display the waveform of a varying electrical signal. The heart of an oscilloscope is its cathode-ray tube. The cathode-ray tube consists of a specially shaped glass vacuum tube containing a number of electrodes. The first of these is the cylindrical cathode which is heated by a filament powered from a low-voltage supply. When the cathode is hot it emits electrons, which form an electron cloud around the cathode. But, when the extra-high-tension power supply is switched on, the electrons are accelerated toward the other electrodes which are held positive with respect to the cathode. These electrodes are the grid, a focusing anode and a main anode. The electrodes are cylindrical so the beam passes through them and strikes the tube face. This is coated with a phosphor which glows when hit by electrons, the number of electrons that hit the tube face (or screen) in a given interval determines the brightness of the glow.

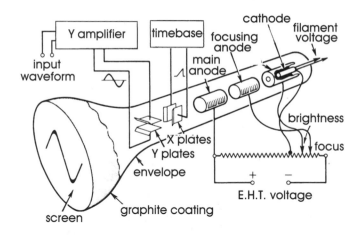

The main parts of a cathode-ray oscilloscope.

The different potentials of the electrodes are obtained by tapping various points on a resistor. The main anode is positive with respect to the cathode, to provide the basic accelerating force. The grid is slightly negative of the cathode, removing some of the electrons from the beam. This acts to control the number of electrons reaching the screen and hence the brightness of the spot. The focusing anode is also negative with respect to the cathode. The focusing electrode is shaped so that it creates an electric field, which focuses the beam to a fine spot. This potential too can be adjusted to focus the beam precisely.

After passing through the main anode, the electrons are travelling at very high velocity. After impinging on the screen they eventually find their way back to the cathode by way of a graphite coating on the inner surface of the tube. As they strike the screen they give up energy to the phosphor to make it glow brightly. If the electron beam is stationary the tube will have a stationary bright spot in the centre of the screen. The colour of the spot may be green, blue, orange, or some other colour, and it may have long or short persistence, depending on the composition of the phosphor. Different phosphors are used depending upon the purpose of the tube. The term 'phosphor' has nothing to do with the element phosphorus. It is used to describe any substance used for coating the screen simply because it glows or phosphoresces when activated by the electron beam. The term *persistence* is self explanatory. A long-persistence phosphor may glow for several seconds or more after bombardment. A short-persistence phosphor ceases to glow almost as soon as the beam is cut off. Most tubes used in oscilloscopes have a short-persistence phosphor, so that the rapidly moving beam draws a clear-cut picture on the screen. By contrast a tube used in a radar PPI display (p. 293) needs a long-persistence phosphor so that the images of objects located by the radar remain glowing on the screen until the next revolution of the antenna.

X and Y plates

Between the main anode and the screen the electron beam passes between two sets of deflector plates. The pair known as the X plates deflect the beam to left or right (in the x-direction on a conventionally plotted x-y graph). A timebase circuit produces a sawtooth p.d. which is applied between the plates to repeatedly deflect the beam steadily across the screen from left to right and then rapidly flick it back to the left. The effect of this is to produce a horizontal line across the centre of the screen. Given the right amount of persistence and a suitable repetition rate, the line appears to be constant.

The Y plates are situated above and below the beam and deflect it in a vertical direction. The alternating p.d. that is to be observed on the screen is fed to an amplifier, the Y-amplifier, which produces a corresponding alternating p.d. between the Y plates. The action of the sawtooth generator is synchronized with the input signal so that it repeats for each cycle of the incoming wave. Given the correct adjustments of the circuits, as the wave goes through one cycle the spot of light moves from left to right across the screen. Instead of drawing a straight horizontal line, the beam draws a graph of the waveform. In the figure we show a sine-wave input and the resulting sine wave displayed on the screen (except that in reality the display would be a bright line on a dark ground).

Given other waveforms such as square waves, triangular waves or the complex waveforms of audio signals, the corresponding shape is displayed on the screen.

The screen of an oscilloscope is usually marked with a one-centimetre square grid to allow input voltages to be measured on the screen. The scale of the display can be set in both directions. The rate at which the beam scans horizontally across the screen can be set to a range of values so that 1 cm in the horizontal direction corresponds to 1 s, 100 ms, 10 ms, down to 0.1 μs. The shorter times allow signals in the radio-frequency bands to be displayed. The gain of the Y-amplifier can also be set so that 1 cm in the vertical direction corresponds to 20 V, 100 mV, 10 mV, down to 1 μV.

The paragraphs above describe the simplest form of oscilloscope, but other more elaborate models are available. Oscilloscopes often have a second beam with its own timebase and amplifier so that two signals may be observed at the same time and compared. There are also triple-beam instruments. Another refinement is trace storage, in which a single cycle of a waveform (which might be a single pulse of complicated shape) is captured and stored, then displayed on the screen to allow its features to be examined.

With the increased popularity of personal computers another approach has been made to observing waveforms. Much of the function of the cathode ray tube (hardware) can be taken over by a computer program (software). The signal to be examined is fed to an amplifier which is interfaced to the computer. The computer samples the signal at regular intervals, perhaps several million times per second and converts each voltage sample into its corresponding binary form. This data is then processed in the computer in a variety of ways. It can generate on its screen a display identical in appearance with that of the cathode ray tube. Also there are many other ways in which the data can be stored and analysed by the computer to obtain other information from it. There are also more elaborate multimeters that are able to plot a waveform on a small built-in LCD screen. These 'graphic multimeters' lack the definition of a CRO but the feature is a useful extension of the functions of a multimeter and the instrument is much more portable than a CRO.

Signal and pulse generators

The simplest sort of signal generator is an oscillator, usually based on just one IC enclosed in a matchbox-sized case with a metal probe at one end. The frequency is about 1 kHz and its function is to inject a signal into a test circuit.

Using an oscilloscope, a headset or the audio output stage (if it has one) of the test circuit we can check that the signal is passing along from stage to stage in the circuit. If we find that it reaches a certain point but is not detected beyond that point, we know where to begin to look for faults.

Signal generators are also used for investigating the behaviour of circuits over a range of audio or radio frequencies. The amplitude and the frequency of the signal can be switched over several ranges. Frequency can be set from a few cycles per second up to several hundred megahertz. Usually a selection of waveforms is available including sine, square, triangular and sawtooth waves. A pulse generator is able to produce nominally square pulses of known duration and with specified rise times and fall times. The interval between pulses may also be adjusted.

Logical testgear

With the increased swing toward digital circuitry, many new digital test instruments have been devised. One of the simplest is the **logic probe**. This is about the size of a thick writing pen and there is a lead by which it takes its power from the power rails of the test circuit. It ends in a metallic contact rod that can be pressed against the terminal pins of logic circuits to ascertain their logic state. This is indicated by a bicoloured LED which glows red or green, depending on whether the tested point is at logic high or logic low.

At the other extreme are the logic analysers that display the signals on several channels simultaneously in a similar manner to a CRO. Usually the display device is a graphics LCD. Input logic signals from up to 32 inputs are stored in memory at rates of several tens of megahertz. These can be played back under control so that the engineer can examine the logic states and the sequence in which they occur. Particular combinations of logic signals can be searched for and identified. The stored data can also be printed out for record and analysis.

Finally, there are the logic IC testers. These are usually hand-held devices with a socket to accept the IC. They are programmed to apply a set sequence of inputs to a given type of logic IC and to confirm that the expected outputs are produced at each stage. These testers can also identify the type of an unknown logic device.

I n the early days of electronics, a circuit was built by bolting or screwing the components to a chassis of sheet aluminium or wood and then joining their terminals by stout cotton-covered copper wire. A gradual reduction in the average size of components, plus the need for mass production has led to this technique being totally replaced by those described in this chapter.

Breadboard

A breadboard (sometimes called a *plugblock*) is used for building temporary circuits. It is useful to designers because it allows components to be removed and replaced easily. It is useful to the person who wants to build a circuit to demonstrate its action, then to reuse the components in another circuit.

A breadboard consists of plastic block holding a matrix of electrical sockets of a size suitable for gripping thin connecting wire, component wires or the pins of transistors and integrated circuits (ICs). The sockets are connected inside the board, usually in rows of five sockets. A row of five connected sockets is filled in at the top right of the figure. The rows are 2.54 mm apart and the sockets spaced 2.54 mm apart in the rows, which is the correct spacing for the pins of ICs and many other components.

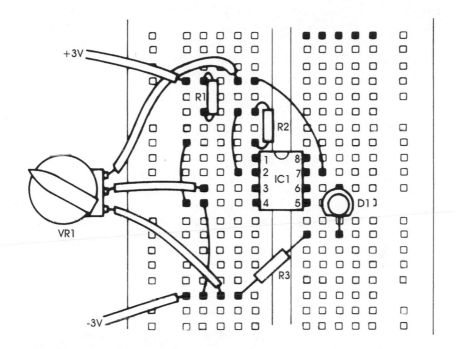

A breadboad has rows of sockets, connected in groups.

On some designs of board, longer rows of sockets occur along the edges of the board, usable for power supply rails. On the board illustrated above, the sockets on the extreme right, although spaced in vertical groups of five are all connected together. The figure shows how a simple circuit is built up on the board. For example, the +3 V supply is connected to R1, to VR1 and (through a wire link) to pin 7 of IC1. Pin 6 of IC1 is connected to the LED (D1), which is connected through R3 to the –3 V supply. Note the gap in the centre of the board, bridged by the IC. This ensures that opposite pins of the ICs are not connected.

Once a circuit is assembled on the breadboard, it is tested. If its performance is found to be less than perfect, it is easy to substitute resistors or capacitors of different values. It is also easy to replace components suspected of being faulty. If sections of the circuit need to be isolated to investigate faulty operation, this may readily be done by removing one of the component terminal wires from its socket.

Breadboards are ideal for building and testing relatively simple circuits. Although they can, in theory, be used for complicated circuits, the board soon becomes covered with a nest of wires so that it becomes very difficult to trace the connections. If one of the wires is accidentally removed from its socket, it is often difficult to find the correct socket in which to replace it.

Stripboard

Stripboard is one of the commonly-used types of prototyping board. These boards are intended for permanently assembling one-off circuits, especially prototypes. The board is made from insulating material, usually a resin-bonded plastic or fibreglass. One side has parallel copper strips on it, spaced 2.54 mm apart. There are holes bored in these strips, also 2.54 mm apart. Components are placed on the other side of the board with their wires bent to pass through the holes. The wires are soldered to the copper strips, the projecting ends being cut off to make the assembly neater.

The strips correspond to the rows of sockets on breadboards, allowing several components to be joined together, but a strip has many more than five holes so numerous connections are possible. Using a special tool, strips can be cut into shorter lengths where it is necessary to use one strip for making several different connections. A circuit built on stripboard has permanently soldered connections, but it is not difficult to remove a component and solder in one of a different value, or even to completely modify part of the circuit. A prototype circuit may be built and tested stage by stage.

A portion of stripboard, as seen from the strip side. Four cut strips are visible on the left. Component wires have been pushed through the holes from the front of the board, soldered to the strips, and then cut short.

There are various specialized designs of stripboard, some intended for use with integrated circuits. Others have a built-in connector pad at one end. Specially shaped and laid out prototype boards are available for assembling prototype cards for computers.

Printed circuit board

A printed circuit board (or PCB) begins as a board of resin-bonded plastic or fibreglass coated on one or both sides with a continuous layer of copper. It is then etched (see below) to produce a circuit layout, with pads to which components are soldered, and tracks to provide the required connections between the pads. Holes are drilled in the pads to accept component leads or pins and occasional wire links. During assembly all that is necessary is to insert the leads or pins into the holes, solder them to the pads and trim the surplus leads. There is no 'wiring up' of the board, other than connecting it to off-board components, as the wiring stage is taken care of when the layout is designed.

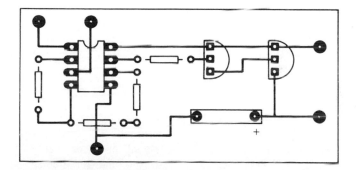

Printed circuit board design, showing square, rectangular and circular pads joined by tracks. The outlines of components, which are mounted on the opposite side of the board, are drawn in thin lines. The components comprise an 8-pin IC, an electrolytic capacitor, two transistors and four resistors. There are five terminal pins for off-board connections.

PCBs may be single-sided or double-sided, but the components are usually mounted on only one side, to make assembly easier. Connections between tracks on opposite surfaces of the board are made where necessary by drilling holes and soldering a double-headed rivet-like **via** to the matching pads on each surface.

The first stage in manufacturing a PCB is to design the layout. Special computer programs are available to assist with this (see title photograph), though a simple circuit can well be designed by hand. The aim is to arrange the components so the connections between them are as short as possible and that there are as few places as possible at which tracks need to cross each other. Naturally, tracks can not actually cross. Crossings can often be avoided by running tracks beneath components. Two examples can be seen on the layout on the previous page, where tracks pass underneath the IC and the resistor. Where a crossing is unavoidable, we use a wire link or, in the case of a double-sided board, route the track on to the other surface. When the design is complete, it must be thoroughly checked because it is difficult to correct errors at the later stages.

The layout is transferred to the board by printing its design in etch-resistant ink. This has the big advantage that thousands of identical boards can be manufactured using the same printing plate. The board is then immersed in a bath of etchant solution. This removes the copper from all areas of the board except those covered by the resistant ink which is known as **resist**. At the end of etching, we are left with the pads connected by tracks. The resist is removed by a second bath. Usually the copper is tinned to protect it from corrosion and to make it easier to solder to. Finally, the board may have a lacquer layer printed on all over except for the pads, as corrosion protection.

Next, the pads are drilled to accept the wire or pins of the components. The spacing of the holes is standardized so that components of standard size can be dropped by machine into the holes. Often special short-leaded resistors are used to make it easier to insert the wires automatically and to eliminate the need for clipping off the excess wire later. The components may be soldered in place by hand, but more often the board is passed through a bath of molten solder in which a 'wave' of solder sweeps across the board, soldering all junctions as it flows.

This is a technique specially suited to mass-production of items such as TV and radio sets and computers, and the description above outlines the production-line stages. The primary advantage of PCB manufacture, in addition to the high rate of circuit production, is that it does not require skilled operators to hand-wire the connections between components.

Hand-wiring is always subject to a small percentage of mis-connections. If the original PCB layout is correct, it can be guaranteed that every board will be correctly 'wired'.

Although the technique was primarily developed for mass production, the PCB has found favour with the home constructor. Many electronics magazines sell PCBs of the projects which they have published, and most electronics kits include a PCB. Errors of wiring are eliminated, but it is still necessary to make sure that all the components are correctly inserted and that the solder connections are properly made. The home constructor can make one-off PCBs using one of several techniques to apply the resist. The simplest method is to draw the tracks by hand using a special lacquer pen. A neater method is to use rub-down pads and tracks, which are similar in use to rub-down lettering.

A popular method is to use copper-clad boards which are coated with a layer of photoresist. A positive transparency of the layout design is placed on the surface, which is exposed to UV light. Positive transparencies are obtained by hand-drawing on film, or using rub-down pads and tracks on film, or by photocopying on film from the printed designs published in books or magazine. The exposed board is then developed, rather like a photographic print, so that the photoresist is removed except where the tracks and pads are to be. After the resist pattern has been created on the board, the rest of the procedure is much like that in the industrial process, except that the home constructor is more likely to solder the connections individually and not use a solder bath.

To sum up, PCBs offer reliable mass production and freedom from wiring mistakes. The only disadvantage, which is not usually important in large-scale production of successful designs, is that it is difficult to alter or modify the circuit once it has been committed to a PCB.

Integrated circuits

On page 81, we described how a transistor is fabricated on a slice of silicon. Resistors, capacitors and other components can be made in a similar way, as well as the connections between them so that a complete circuit can be produced on a single chip. This is known as an **integrated circuit** or IC.

The figure overleaf shows how the resistors of an IC are made. The resistance of a volume of semiconductor depends on its length, its thickness, its width, and on the resistivity of the semiconducting material. The resistivity of semiconductors is high relative to that of metals. This means that even a microscopically small resistor can have a suitably high resistance.

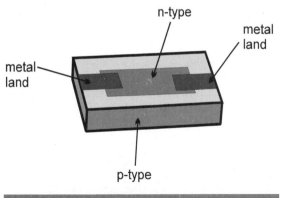

A solid-state resistor is made on a p-type silicon chip by doping a surface region to convert it to n-type.

To make the resistor, a volume of semiconductor of specified dimensions is doped, using the masking techniques described earlier. The resistivity depends upon the amount of doping. Using another mask, areas of metal, known as **lands** are deposited on the chip to act as contacts. The lands may be extended to make contact between the resistor and adjacent components.

The figure shows a resistor of n-type material formed in a slice of p-type material. Conduction is possible from the n-type to the p-type so the resistor is not electrically isolated from the substrate (the p-type). This could cause problems when several resistors and other components are fabricated close together on the same chip. The answer to this is to bias the substrate to be at a lower potential than the resistor. This, in effect, makes the resistor-substrate junction a reverse-biased p-n junction, preventing the flow of current between resistor and substrate. Resistors can also be made of p-type material on an n-type substrate.

Crystals for ICs

ICs are produced on silicon wafers cut from a large silicon crystal. The diameter of a standard silicon crystal is 8 inches, but there is a trend to using larger crystals 12 inches or more in diameter.

To make a capacitor on an IC, we rely on the principle of the varicap diode (p. 79). A layer of p-type material is diffused into an n-type substrate and then a layer of n-type is diffused on top. Metal contacts are deposited at A and B. The diode is reverse-biased, creating a depletion layer which acts as the dielectric.

When we use a varicap diode, the idea is to vary the reverse bias to vary the width of the depletion layer and thus obtain a variable capacitor. In an IC, it is more usual to have a constant bias voltage and to produce capacitors of different fixed values, depending on the dimensions of the plates. Capacitors may also be built up on a silicon block, using a layer of silicon dioxide as the dielectric.

It is virtually impossible to fabricate an inductor in semiconductor material, but the functions of inductors can often be accomplished by sub-circuits built from an amplifier and a few capacitors and resistors (see p. 201).

Integration

The way in which the basic components described in the previous section may be integrated into a single on-chip circuit is illustrated by the amplifier shown below.

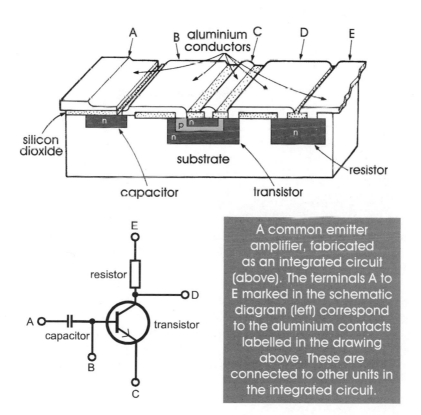

A common emitter amplifier, fabricated as an integrated circuit (above). The terminals A to E marked in the schematic diagram (left) correspond to the aluminium contacts labelled in the drawing above. These are connected to other units in the integrated circuit.

The capacitor is formed by the aluminium contact A acting as one plate, a layer of silicon dioxide forming the dielectric and a layer of n-type material acting as the other plate. The aluminium contact B connects the capacitor to the base (the p-type region) of the transistor. One or two resistors (not shown, for simplicity) would be connected to contact B to bias the transistor into conduction. In the transistor, the n-type layer nearer the surface is the emitter, and has a metal contact C for connection to the 0 V supply rail. From the lower n-type layer, the collector, there is a metal contact D from which the output signal of the amplifier is taken. This connects through a resistor to the positive supply at contact E.

IC manufacture

Integrated circuits may require several consecutive stages of masking etching and doping to build up the complex and interconnected semiconductor devices on the wafer. It may become difficult to control the depth of doping accurately and to prevent dopant from spreading to areas where it should not be. In such cases an ion gun, accelerating ions of the dopant at high speed in a vacuum, is used to impregnate selected areas of the wafer with dopant. This is known as **ion beam doping**. Another technique which may be used at certain stages in production is growing a layer of n-type or p-type silicon on the surface of the wafer by condensing it from a vapour at high temperature. This is rather like the formation of crystals of hoar frost on a window or on the branches of a bush from air saturated with water vapour. This is known as **epitaxial growth**.

The final step in the manufacture of ICs is to test the individual circuits on the wafer, marking and discarding those which are defective. They are separated and each is mounted on a support. The entire amplifier like that in the figure occupies an area of silicon less than 0.1 mm square. Many more complicated ICs are only 1 or 2 mm square. It is not easy to connect these directly to an external circuit. The silicon chip is therefore mounted in a standard package, usually made of plastic but occasionally of ceramic materials or metal. The plastic **double-in-line** (DIL) package is the most commonly-used type. Within the package is a framework of conductors leading to the terminal pins. The circuit chip is mounted in the centre. Terminal lands on the chip are connected to the terminal pins by thin gold wires soldered at either end.

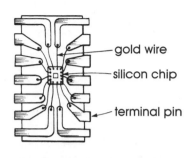

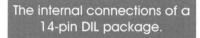

gold wire

silicon chip

terminal pin

The internal connections of a 14-pin DIL package.

A 14-pin IC containing complicated logic circuits.

Special machines are used to perform the soldering quickly yet precisely. The package is then sealed.

DIL packages may vary in their number of terminals from 6 pins to 40 pins, but the majority have 14 pins or 16 pins and look monotonously alike. Yet the variety of type numbers printed on the packages testifies to the wide range of types available. Other packages are often used. Some of the simpler ICs, such as bandgap ICs (p. 108) need only two or three terminal connections. These can be contained in a plastic package identical with that used for transistors. At the other extreme, a microprocessor IC may have over 200 terminal pins. The package is square with four rows of pins along each side. The different types of IC are outlined in the next section.

Applications of integration

The main types of IC fall under the following broad headings;

Linear ICs These include amplifiers of many kinds. Some are specially designed for audio applications and many of these operate at high power. Those of the highest power, often over 200 W, are provided with built-in heat sinks to dissipate excess heat. ICs for use in stereophonic audio systems have two identical amplifiers on the same chip, one for each channel. Radio-frequency amplifiers are also available in integrated form. One of the most versatile of amplifiers is the **operational amplifier,** or op amp. It is a high-gain amplifier with special features that make it useful in many applications, including audio amplification and instrumentation. The ways of using op amps are described in Chapter 18.

Logic ICs The action of these is described in Chapter 19. They are also known as **digital ICs.** The early logic ICs provided only the simplest of functions such as the elementary logical operations NAND and NOR and required only a dozen or so integrated components on one chip. These ICs were followed by circuits of greater complexity such as counters, data storage registers, and arithmetic devices that could add two four-digit binary numbers. This level of complexity is known as **medium scale integration** (or MSI). As IC manufacturing techniques improved, and it became possible to fabricate the devices with smaller and smaller dimensions and so cram more and more on to a chip, MSI was followed by **large scale integration** (LSI) and later by **very large scale integration** (VLSI). Whole logical systems are fabricated on a single chip.

A pocket calculator and a digital clock or watch have their complete logical circuit on a single LSI chip. The heart of the microcomputer is a single microprocessor chip in which all the logical and mathematical operations of the computer are performed. In the Pentium microprocessor chip there are over 1 million transistors and almost 20 million p-n junctions, a good example of VLSI. Other examples are the **microcontroller** chips which comprise not only a microprocessor but also the control logic and memory, so that they are a complete computer on one chip. They can be manufactured so cheaply that it is economical to incorporate them in automatic equipment such as washing machines, industrial robots, microwave ovens, and automobile and aircraft control systems, even though one particular application does not make full use of all the microprocessor's abilities. Most personal computer systems have a microprocessor to operate the video monitor and another to operate the printer as well as the main central processing unit (CPU) in the computer itself.

Special function ICs It is impossible to cover the wide range of ICs that come under this heading. We have timer ICs that, with a few external components, can control the duration of time periods from a few microseconds to several days or even months. We have the phase-locked loop ICs that can be tuned to pick out and lock on to a signal of a particular frequency from a mass of other signals. There is the radio IC looking just like a transistor that, with a tuning capacitor plus two or three resistors, gives a complete radio set. There are the remote-control ICs that automatically generate a coded stream of pulses: when this code is received by the TV set, another IC decodes the message and controls changing of wavebands, the brightness and colour balance of the picture or the volume of the sound. There are special ICs for producing the many voices and rhythms of the electronic keyboard and synthesizer. There seems to be no limit to the possibilities of integrated circuits. Whenever there is a new model of a device, such as mobile telephone or a digital camera, it is usually most economical to design a new special function IC to control it.

Gallium arsenide ICs

Although silicon is the most widely used base for integrated circuits, an alternative material is gallium arsenide. Devices based on this compound are much better for high-speed applications, including fast logic gates. They also use less power. Apart from the special safety precautions that must be taken because of the arsenic in the material, galllum arsenide has the further disadvantage that it is more difficult to work with, raising the cost of manufacture. In particular it is more difficult to produce an insulating oxide layer on gallium arsenide than it is on silicon. High temperatures are necessary, which may impair the structure of the crystal. A recent development is to grow a layer of nitride instead, a technique which is easier to carry out.

Advantages of integration

The development of integration techniques has brought about a major revolution in electronics and consequently in many aspects of our lives. The major advantages of integrated circuits are:

- Miniaturization — circuit boards and therefore electronic equipment can be smaller.
- Low cost — the simpler ICs cost no more than a single transistor.
- Production of circuits using ICs is more easily automated, leading to lower costs.
- Use less power (especially CMOS, p. 212); an advantage for portable equipment.
- Greater reliability.
- Longer life.
- Easier servicing — instead of testing individual components, simply replace the IC.
- Small size of transistors and the very small distances between components on the chip mean that the circuits are faster-acting and can be operated at much higher frequencies.
- Components that are close together on the same chip are all at the same temperature; this improves the stability and precision of certain types of circuit.

Surface mount technology

Surface mount resistors are described in Chapter 3. Most of the conventional semiconductor devices are also available as SMDs. The photograph below shows a typical transistor package. There are no special problems in producing this as the actual transistor is small (p. 81) and only needs putting into a smaller package. The terminal pins are shaped to make even contact with the surface of the board. Diodes and LEDs are produced in the same package, the LEDs being moulded in coloured transparent plastic. ICs too are basically very small so they can readily be fitted into a smaller package half the conventional size, with terminal pins spaced only half the usual distance apart.

An assortment of SMDs on the circuit board of the Stamp® stamp-sized computer. Two of the ICs are 8-pin DIL types, but its pins have only half the normal spacing and are soldered to the upper surface of the board. Between them is a transistor in the standard SOT23 package. On the right is a 20-pin PIC controller IC, with pins at only a quarter of the normal spacing. The complete system fits on a circuit board only 11 mm × 37 mm.

PCBs are prepared by masking and etching as for conventional through-hole construction with the important difference that no holes are drilled. The shapes and spacing of the pads are commensurate with SMD terminal sizes and spacing.

Another important difference is that components may be mounted on both sides of the board, thus effectively doubling the number of components that may be accommodated on a board of given area. Any connections required between the two sides are made by carrying short leads around the edges. Before soldering, the board is printed with a pattern of solder paste so that there is a blob of paste on each pad. The paste consists of a mixture of minutely powdered solder and a soft resin. It is adhesive so that, when the components are placed on the board (either by hand or by machine) they remain in place. Blobs of adhesive may also be put on the board to hold the larger components such as electrolytic capacitors in place. In industrial production, all components are soldered at once using a blast of hot air, or by passing the board through an oven at a controlled temperature. The solder melts and flows into the joints. On the small scale, the solder is melted by touching the joint with a hot soldering iron.

SMT is more suited to the equipment manufacturer rather than to the home constructor, but several SMT projects have appeared in magazines. It is said that small is beautiful, and there is fun to be had in working with such tiny components. The main requirements are a pair of forceps, a powerful magnifier, good eyesight, and a modicum of patience.

Computer simulations

Circuit design is something of an art. Although it is possible to predict the behaviour of a very simple circuit mathematically, there are so many factors to consider in a more complicated circuit that the calculations become impossibly convoluted. This is where the breadboard and subsequently the stripboard are so useful. Having decided on initial values for components, the final values may be arrived at by a process of trial and error.

Computer software is able to simulate the action of components, using mathematical routines (called **models**) that would take far too long to perform manually or with an ordinary calculator. It can take the signals from one component and feed them to another, so simulating the action of a complete circuit. The intended circuit design is keyed in to the computer as a **netlist**. This lists the components, their values and other characteristics, and the connections to be made between them. Most simulation programs have a facility by which, instead of keying in a netlist, the user draws the circuit diagram on screen, typing the values and other characteristics of the components on the diagram. The program uses this diagram to prepare a netlist automatically. The software is then asked to perform analyses of the behaviour of the circuit.

These can be simple analyses which, for example, produce a table of the potentials at all points in the circuit network; or they can be far more complicated, for example, showing the output waveform of an amplifier when a given signal is supplied to it. The figures on pages 63 and 64 which demonstrated the action of a rectifier circuit, were obtained from a simulation, using the SpiceAge© software. The action of the software is the same as having a real circuit on a breadboard and a range of instruments such as voltmeters, signal generators and oscilloscopes with which to test it. Changing component values or connections is only a matter of quickly editing the netlist or diagram; there is no need to unsolder a component and solder in a different one. And, of course, software 'components' are never faulty, and can never become damaged or burnt out.

Transistor and IC manufacturers supply libraries of 'components' in data form which match the behaviour of their real components, so that it is possible to specify exactly which type number is to be used in the circuit. It is also possible to 'sweep' component values (such as resistance, capacitance, and gain) over their full tolerance range so that one can be certain that a circuit will work whatever the actual value of any component used in the real circuit. Similarly, the operating temperature can be swept to make certain that the circuit works equally well whatever the ambient temperature.

Many programs have been written for circuit simulation but most are variants and enhancements of a program known as SPICE. This is short for Simulation Program with Integrated Circuit Emphasis, pioneered and developed at the University of California, Berkeley. It was intended for designing ICs, which by their nature are not amenable to breadboarding, but has subsequently been extended to use with discrete components.

Using simulation software saves the designer considerable time, and it virtually eliminates the need for breadboarding. The design is perfected on the computer and is immediately ready to transfer to the PCB.

16 Oscillating circuits

The output voltage of an oscillating circuit alternates between two peak values at regular intervals of time. It may switch sharply between the two extreme values, in which case we call it a **square-wave oscillator**, but the output waveform may have other regular shapes such as sawtooth, triangular or sine waves. The time taken for one complete oscillation or cycle is known as the **period** of the oscillation. The rate of repetition of the waveform is known as the **frequency**. Frequency is expressed in hertz (symbol, Hz) where 1 Hz is equal to 1 cycle per second. Period and frequency are related by the equation:

$$\text{frequency} = \frac{1}{\text{period}}$$

For example, if the period is 1 ms, the frequency is 1 kHz.

In most, but not all, of the oscillators described in this chapter the timing of one cycle, and hence the frequency of the oscillations, is determined by the time taken to charge and discharge a capacitor through a resistor. The first example illustrates this point.

Bistable circuit

This is not an oscillator but an explanation of how it works will help to explain the action of one of the oscillators described later. The word *bistable* is applied to a circuit that is stable in either one of two states. There are several ways of building a bistable. One version comprises two transistor switches, cross-connected so that the output of one switch becomes the input of the other. The outputs and inputs are not connected directly but through resistors R2 and R3.

When the power is switched on, the circuit assumes one of its stable states. Assume that it begins with the transistor Q1 switched on. If so, output A is at low potential and the base of Q2 is at low potential. This means that Q2 is off. Output B is therefore close to the supply voltage (call this 'high') and the base of Q1 is receiving enough current to keep Q1 switched on. The circuit is stable in this state and remains in this state indefinitely. Note that output A is low and output B is high.

Now suppose that we connect input A (the base of Q1) to the 0 V rail or in some other way give it a low input. Immediately we do this, Q1 is turned off. Output A becomes high. This turns Q2 on and output B goes low. Even though we may connect input A to the 0 V line only for an instant, it is now permanently low and the circuit is stable in this state. Note that output A is high and output B is low. In either state, one output is high and the other one is low. It is not possible for both to be high or both to be low.

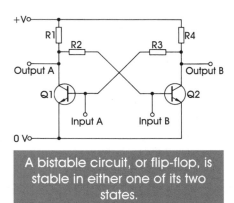

A bistable circuit, or flip-flop, is stable in either one of its two states.

Connecting input A briefly to the 0 V line has made output A high and output B low. Connecting input A to the 0 V line *again* has no futher effect, as the base of Q1 is already at low potential. But see what happens if we connect the *other* input (B) briefly to 0 V. This turns off Q2, output B goes high; Q1 is turned on, output A goes low. The circuit has returned to its original state.

As we connect the inputs A and B to 0 V alternately, the circuit switches between one stable state and the other. It 'flips' and it 'flops', which is why this type of circuit is also called a **flip-flop**.

If we connect up this circuit and leave it for someone to play with, we can always tell which input was most recently grounded. The circuit 'remembers' which input was most recently connected to 0 V. So, although this circuit is not an oscillator, it is very useful. Flip-flops are a way of storing data, and are used in the memory circuits of computers.

Monostable circuit

The term 'monostable' implies that this circuit is stable in only one state. Like the bistable, it has two states but it is stable in only one of them. When it is put into the unstable state, it sooner or later returns to its stable state. The circuit is very similar to that of the bistable. It consists of two cross-connected transistor switches but, instead of linking them both through resistors, only one link is a resistor and the other link is a capacitor. This makes time a factor in the operation of the circuit.

When the power is switched on, the circuit goes into its stable state with Q1 off and Q2 on. The collector of Q2 (and also the output) is therefore low. No current flows to the base of Q1, so Q1 is off. Now consider the p.d. across the capacitor. Plate X is high, almost at the supply voltage. Plate Y is at about 0.6 V, the typical base-emitter voltage of a transistor.

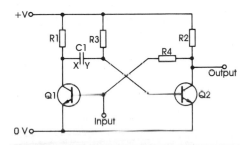

A monostable circuit, or pulse generator, is stable in only one of its two states.

If the supply voltage is 6 V, the p.d. across the capacitor is about 5.4 V, with X positive of Y. Now a brief positive pulse is applied to the input, for example, by touching the input to the +V rail. This turns Q1 on and the potential of plate X falls rapidly to (say) 1 V. This is a drop in potential of about 5 V. The capacitor acts to keep the p.d. across it constant. If X drops from 6 V to 1 V, the potential of Y drops by 5 V from 0.6 V to –4.4 V. This turns Q2 off. The output goes high instantly.

With plate Y at –4.4 V, current flows through R3, gradually charging the capacitor. The potential of plate Y (and also the potential at the base of Q2) gradually rises. The rate of rise depends on the value of R3 and of the capacitor C1.

Eventually, the potential of Y reaches 0.6 V and then current begins to flow into Q2, turning it on. As soon as it starts to turn on, the output falls to low, and this turns Q1 off. Its collector potential goes high, raising the potential of plate X almost to the supply level again. This forces the potential of plate Y up, completely turning on Q2. While the capacitor was charging, the circuit was unstable, but now it is stable again.

Summing up, the action of this circuit is that a brief high pulse to the input causes the output to go high for a period of time depending on the values of R3 and C1. In a practical monostable of this type, the output may be high for several seconds or even minutes. This circuit is not an oscillator but has many applications as a pulse stretcher (converting a short pulse into a much longer one) and as a delay circuit.

Astable circuit

The word 'astable' means '*not* stable'. This circuit is not stable in either state; it switches continually from one state to the other. It oscillates. The circuit is very similar to that of the bistable and monostable, the difference being that the connections between the transistor switches are *both* made by way of capacitors. As in the other two circuits, at any instant one transistor is on and the other is off.

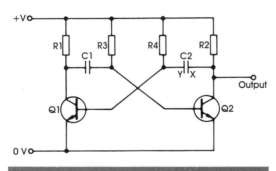

An astable circuit is not stable in either of its two states, but continually oscillates between one state and the other.

We begin with Q1 off and Q2 on. Output is low. If Q2 has only just been turned on, the output has only just gone low and the potential of plate X of C2 has fallen sharply. As in the previous description, this takes the potential of plate Y and that at the base of Q1 to a negative value. Q1 is firmly off, but it does not remain in this state.

Current flows though R4, gradually recharging plate Y. When the potential at Y and the base of Q1 reaches 0.6 V, Q1 begins to turn on. The potential at its collector starts to fall; this pulls down the potential on both plates of C1, making the base of Q2 negative and turning it off.

Now the circuit is in its other state, but it does not remain in this state either. Current flows through R3, gradually raising the potential at the base of Q2 until it turns on again. The circuit returns to its original state.

This circuit automatically oscillates between its two unstable states at a rate determined by the values of R3, R4, C1 and C2. If the values are such that the circuit oscillates at a few hundred or thousand hertz, it can be used to generate audio-frequency signals. If it oscillates more slowly, it can be used for flashing warning lamps.

The 555 timer IC

This must be one of the most useful ICs ever invented. The figure shows it being used as the basis of a monostable circuit. As can be seen, the circuit requires only three components instead of the seven components of the BJT monostable. This makes it much simpler to wire up. Also, because of its design, it gives more precise timing intervals than the BJT monostable, and (in its CMOS version, the 7555) can be used to produce output pulses as short as a few microseconds up to as long as several hours. Once again, the timing relies on the charging of a capacitor through a resistor and the pulse length is set by choosing suitable values for R1 and C1.

When power is switched on, current flows through R1 and begins to charge C1. A circuit inside the IC, and connected to the threshold pin, detects the p.d. across the capacitor. When this has risen to one third that of the supply voltage (*V*/3), the current is then diverted into the discharge pin and so to ground. The capacitor remains charged to *V*/3, and the output is 0 V. The timer is triggered into action by a brief low pulse applied to the trigger pin.

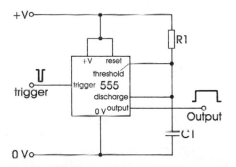

A 555 monostable circuit gives a single positive-going output pulse when it is triggered by a negative-going input pulse.

Instantly, the output goes high. Current no longer flows through the discharge pin, but into the capacitor, increasing the charge on it until it reaches two-thirds of the supply voltage (2*V*/3).

When the internal threshold circuit detects this level, the output goes low. At the same time the discharge pin begins once more to admit current, discharging the capacitor rapidly to $V/3$ again. The output is high for as long as it takes to charge the capacitor from $V/3$ to $2V/3$. Since the *rate* of charging is proportional to V, the time taken to charge is independent of V. This is an important feature of this timer. The supply voltage (within the range 2 V to 18 V for the CMOS version) makes no difference to the length of the output pulse, and it is not affected by the battery being flat.

Used as a monostable, the 555 timer IC has many applications as a process timing, and for producing delays with an accuracy of about ±1%.

The 555 timer IC can also be used in an astable circuit. This requires two resistors. The capacitor charges through both resistors but, when it has charged to $2V/3$, it is discharged through R2 only. The trigger pin is connected to the threshold pin so the IC is re-triggered as the charge falls to $V/3$. The timer is triggered repeatedly, giving a square output waveform.

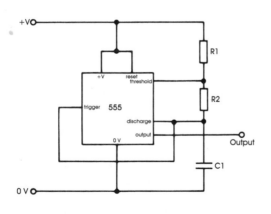

A 555 astable circuit gives a succession of positive output pulses as soon as power is switched on. The output remains continuously low if the reset input is made zero.

The frequency is determined by selecting suitable values for R1, R2 and C1. Since it charges through R1 and R2 but discharges only through R2, the length of the 'high' part of the cycle is longer than that of the 'low' part. One version of this IC can produce frequencies up to 2 MHz, and all versions can run at very low frequencies with cycles lasting several hours. The frequency is unaffected by the supply voltage.

In the astable mode, the 555 timer IC is used in a wide range of timing devices. Operating at audio frequencies, it is often used in tone generators, sirens and audio alarm circuits.

Crystal oscillator

The oscillators described so far in this chapter all depend on charging a capacitor up to a cerain voltage and then letting it discharge. These are known as **relaxation oscillators**. The other class of oscillators, described in this section and the next, are called **resonance oscillators**. These depend for their action on the phenomenon of resonance, described on page 65.

The resonant element is a crystal oscillator, which relies on the mechanical properties of a quartz crystal. Crystals can be cut so as to vibrate at a precisely fixed natural frequency. Electrodes are deposited on two opposite faces. The crystal is made to oscillate by connecting it into a circuit with approximately the same resonant frequency as the crystal. Owing to the piezo-electric effect (p. 118), the oscillations of the circuit make the crystal vibrate at its natural frequency. The p.d.s developed across the crystal force the circuit to oscillate at the same frequency.

One type of crystal oscillator circuit is based on a logical inverter gate. Logic gates are described in Chapter 16 but, for the present, we need only say that, if the input to this gate (on the left) is logic high (say, +5 V), then its output is low (0 V). Conversely, if its input is low, its output is high. The output is fed back to the input though a high-value resistor R1 (10 MΩ). A high output is fed back to give a high input, which immediately makes the output go low. This is fed back, making the output go high again.

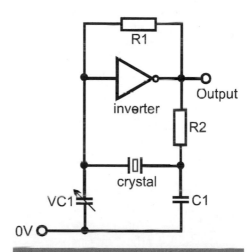

The astable circuit of R1 and the inverter is made to run at a rate that is precisely fixed by the crystal.

The inverter/R1 network has astable properties but, left to itself, it would oscillate at an indeterminate and probably uneven rate. To regulate its oscillations, we take its output, delay it slightly by the R2/C1 network and feed the signal to the crystal. When a positive signal is applied to the crystal, forces within the crystal distort its lattice. When the signal changes to zero, the lattice recovers its shape and delivers stored energy back to the circuit.

As in the previous description of resonance, the situation is like pushing a child on a swing. When we apply a positive signal (= push the child) energy is stored in the crystal (= swing). The energy is returned as a p.d. in the circuit (= the force we would experience if we tried to stop the swinging). If the signal has just the right frequency, matching the natural frequency of vibration of the crystal (= pushing the swing just as it starts to move forward) the circuit oscillates more and more strongly (= higher and higher!). The circuit resonates.

The advantage of crystals is that they do not resonate unless the frequency is very close to the frequency for which they have been prepared. Crystals are therefore used for high-precision oscillators, such as those in clocks and watches. The crystal in a watch usually has a natural frequency of 32.768 kHz (note the five-figure precision). This is a high frequency, which is inevitable since a small object such as a crystal can not have a low natural frequency. The watch has circuits which divide this frequency. At each stage the frequency is halved so that with 15 stages of frequency division the frequency is divided by 2^{15}. This equals 32 768, so the resulting frequency is 1 Hz. One cycle per second is ideal for driving clocks and watches. The precision is such that a fairly inexpensive crystal-controlled clock or watch is expected to be accurate to within 15 seconds per month.

High-frequency crystal oscillators are also used directly to produce the carrier frequency of radio transmitters. The frequency has good stability with temperature and does not vary with the age of the crystal. The high-frequency 'system clock' in a computer, running at a hundred megahertz or more is another example of a crystal oscillator.

Sinewave oscillators

There are many designs of oscillator producing sine waves, each based on a resonant network. In order to introduce the time factor into its operation, the network contains a capacitor, or an inductor, or in some cases both. There are many variations on this theme and here we describe just one of these, the **Colpitts oscillator**.

The resonant network (opposite) consists of an inductor and two capacitors. One of the capacitors is variable so that it can be adjusted to tune the circuit to resonate at any required frequency. The resonant network is placed in the collector circuit of Q1. Oscillations in the collector current cause the circuit to oscillate at the resonant frequency.

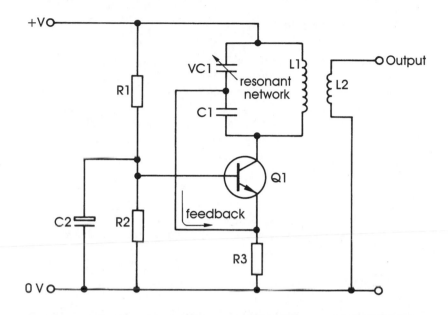

The frequency of a Colpitts sinewave oscillator is set by the values of the capacitors and inductor in its resonant network.

To see how the transistor is made to produce oscillations, we must look at the connections to its other two terminals. The base of the transistor receives its current from a potential-divider, consisting of two resistors R1 and R2. The base potential is held very steady by the action of the large-value capacitor C2. We can consider the base potential as being constant. The emitter of Q1 is connected to a wire running from the junction of the two capacitors in the resonant network. In effect, we are tapping off a small signal from the resonant network and feeding it to the emitter. We call this signal **feedback**. The effect of this feedback is to cause the emitter potential to oscillate at the resonant frequency. With the base potential fixed and the emitter potential oscillating, the base-emitter potential difference is oscillating. The oscillations are only a few millivolts but are amplified by the transistor action, causing a larger oscillation in the collector current. This keeps the circuit resonating. In short, we tap part of the signal from the resonant circuit, amplify it, and use it to maintain the circuit in oscillation. The action is self-sustaining and the circuit oscillates continuously. The output from the circuit is obtained from a second coil L2 wound on the same core as L1. In practice, L1 and L2 make up a transformer. Oscillations in the magnetic field of L1 induce oscillating currents in L2, at the same frequency.

Oscillating signals

Sinewave oscillators, such as the Colpitts oscillator, produce a p.d. which continually reverses in direction. A graph of the p.d. plotted against time (thick continuous line) shows the p.d. starting at zero, then increasing to a maximum value in the positive direction. Having reached its maximum, it decreases to zero again, then begins to increase in the negative direction. After reaching a maximum negative value it decreases to zero again. This completes one cycle, which is then repeated indefinitely. The number of cycles completed in one second is the frequency, in hertz.

The shape of the p.d.–time curve is identical with the curve obtained when we plot a graph of the sine of angles from 0° to 360° (that is, for one cycle, the figure shows two cycles). This is why we call it a sinewave oscillator. If we connect a resistor across the output terminals of this oscillator, the flow of current at any given time is proportional to the p.d., since current = p.d./resistance. The graph of current against time has the same shape as that of p.d. against time. The current flows in one direction then reverses and flows in the other direction, then reverses again, repeating the reversals indefinitely. We say that it is an **alternating current**.

On page 62, we saw what happens if we connect a resistor and capacitor in series across a sine wave oscillator. Now we will look at this more closely. Suppose that we change the resistor to 330 Ω and the capacitor to 470 nF, but keep the amplitude at 1 V and the frequency at 1 kHz, as before. The figure shows how the p.d.s across the oscillator (black curve), the resistor (dark grey) and the capacitor (light grey) vary during the course of three cycles. As before, the p.d across the capacitor is a sine wave of 1 kHz, but with reduced amplitude.

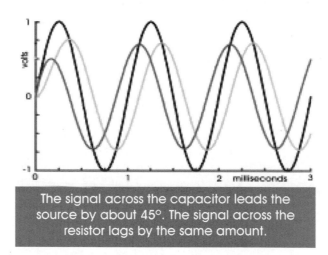

The signal across the capacitor leads the source by about 45°. The signal across the resistor lags by the same amount.

Now look at curve for the resistor p.d. This too is a sine wave and has reduced amplitude, in fact its amplitude is approximately equal to that of the p.d. across the equality. The reason for this is that we have selected values for R and C that give this result at 1 kHz. The impedance of the resistor is 330 Ω and is unaffected by frequency. The impedance of the capacitor is 339 Ω, which is close to that of the resistor. Note also that, because the total p.d. from the source is split into two parts, the resistor p.d. and the capacitor p.d., the resistor and capacitor p.d.s *must* add up to the source p.d. at all times. If you check the graphs, you can see that this is so.

The alternating changes of p.d. across the resistor and capacitor and the alternating current through the resistor follow a complicated but repeating pattern. There are certain important points about this pattern:

- The waveform of all three quantities is a sine wave.

- All three waveforms have the same frequency.

- The resistor p.d. reaches its peaks in either direction slightly earlier than the source p.d.

- The capacitor p.d. reaches its peaks in either direction slightly later than the source p.d.

- The resistor and capacitor p.d.s have almost the same amplitude of about 0.7 V.

The fact that the p.d. sine waves reach their peaks at different times is expressed by saying that they are *out of phase*, or that there are *phase differences* between them. The usual way of expressing a phase difference is to quote the **phase angle**. In the graph, the resistor p.d. reaches its peak about 0.13 ms before the source p.d. in each cycle. If 1 ms (the time of one cycle) is equivalent to 360°, then 0.13 ms is equivalent to $360 \times 0.13 = 47°$. We say that the resistor p.d. has a phase angle of +47°, or we can say that it has a **phase lead** of 47°.

Similarly, the capacitor p.d. reaches its peak about 0.013 ms *after* the source p.d. This is equivalent to a phase angle of –47°. We call this a phase lag of 47°. At this frequency, the two phase differences are equal in size but opposite in direction.

You may have noticed in the figure that the curves during the first cycle do not have quite the same shapes as during the second and subsequent cycles. This is because the capacitor has no charge on it at the beginning of the first cycle. From the second cycle onwards, it already has charge left over from the end of the previous cycle and the waveform is identically shaped in each cycle.

Effects of frequency

We have already seen (p.63) that time is a factor in the charging and discharging of a capacitor. If we increase the frequency of the sinewave oscillator we force the capacitor to charge and discharge in a shorter time, and might expect it to behave differently. The figure on page 64 shows what happens if the frequency is increased tenfold, from 1 kHz to 10 kHz, in the original circuit. A similar effect is observed when R = 330 Ω and C = 470 nF. As before, we show three cycles, but now they take only 300 μs.

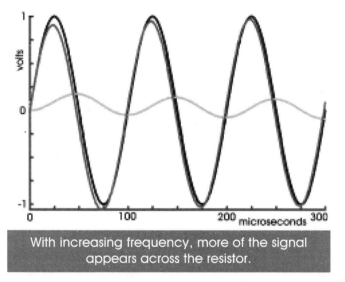

With increasing frequency, more of the signal appears across the resistor.

The same curves are plotted as on page 178, but the pattern they make is very different. The most noticeable effect is that the waveform of the resistor p.d. is almost the same as that of the source p.d. It has almost the same amplitude and only a very small phase lead. By contrast, the capacitor p.d. has a much smaller amplitude than before, and its phase lag has increased to about 90°. This result is almost what we would expect if we replaced the capacitor with a resistor of very low value. Because of Ohm's Law, there could be only a small p.d. across it. The capacitor is acting like a low-value resistance.

If we shift the frequency in the opposite direction, making it ten times less than on page 178 we obtain the result plotted opposite, in which the frequency is 100 Hz. Now it is the curves for source p.d. and the capacitor p.d. which are alike. The amplitude of the capacitor p.d. has increased and its phase lag has almost disappeared. By contrast, the resistor p.d. has reduced in amplitude and its phase lead increased to about 90°. At this low frequency, the effect of the resistor is much less significant than that of the capacitor.

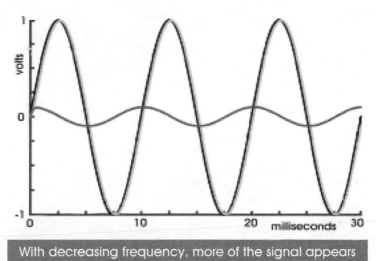

With decreasing frequency, more of the signal appears across the capacitor.

It is as if the capacitor has been replaced by a resistor of very high value, so that currents create a large p.d. across it. The capacitor is acting like a high-value resistance.

Summing up:

•   At high frequencies a capacitor acts like a low-value resistance; high-frequency signals pass easily through it.

•   At low frequencies it acts like a high-value resistance; it blocks the passage of low-frequency signals.

By contrast, an inductor has opposite properties:

•   At high frequencies an inductor acts like a high-value resistance; it blocks the passage of high-frequency signals.

•   At low frequencies it acts like a low-value resistance; low-frequency signals pass easily through it.

If we substitute a 53 mH inductor for the capacitor in the circuit we have just examined, we find that, the p.d.s across the resistor and inductor are equal, as in the figure on page 178. However, the resistor p.d. now lags the source by 45° and the p.d. across the inductor leads it. If we increase the frequency, the inductor p.d. almost equals the source and is in phase with it. Conversely, if we decrease the frequency, the inductor p.d. is much reduced in amplitude and leads the source by up to 90°.

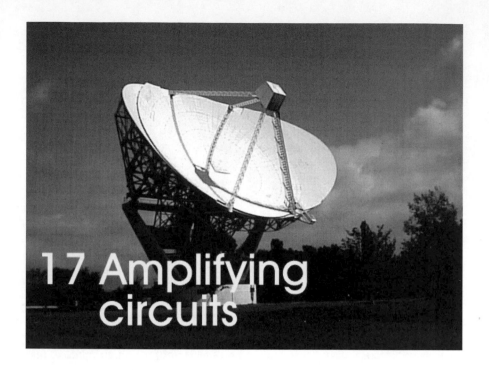

17 Amplifying circuits

In this chapter we describe how transistors together with a few other components are used to build simple amplifier circuits. The amplifying action of a bipolar junction transistor was described on page 82, but now we look at this more closely.

Transistor characteristics

There are various test circuits that are used to measure the way in which a transistor behaves. The results of one of the more useful of such tests is illustrated opposite. The transistor is set up in the common emitter connection as on page 98 with additions to allow the base current and the collector emitter p.d. to be controlled and with meters to measure the base and collector currents. First look at the lowest curve in the figure. The base current I_B is set at 50 μA and the collector-emitter p.d. is gradually increased from 0 V to 30 V. At each stage, we measure the collector current I_C. Collector current begins at zero but rapidly increases to 10 mA. From then on, further increase in the collector-emitter p.d. produces virtually no further increase in I_C. The line is almost horizontal.

If we repeat the trial but make I_B equal to 100 µA (double the previous I_B), the shape of the curve is as before but now levels out with I_C equal to 20 mA (double the previous I_C). The same applies to the other two trials illustrated in the figure. In each test I_C levels out at a value that is 200 times that of I_B. Collector current is proportional to base current, with a **current gain** of 200.

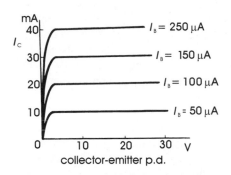

A BJT is a current amplifier.

The figure clearly demonstrates the amplifying action of the transistor. It also shows that, once the collector-emitter p.d. is greater than about 2 V, the amplifying action is independent of the actual collector-emitter p.d. Increases in I_B result in proportionate increases in I_C, but there is a limit to this, not shown in the figure. Once I_B has been increased above a certain value, there is no corresponding increase in I_C. Then we say that the transistor is **saturated**.

For a transistor to amplify as shown in the figure, it must be in its **operating region**. The lower limit of this region is when the collector-emitter p.d. is very low and the transistor is operating on the steeply sloping part of the curve. The upper limit of this region is when I_B is sufficient to saturate the transistor.

Common emitter amplifier

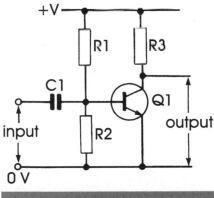

A common emitter amplifier.

On the left is a single-transistor amplifier, using the transistor in the common emitter connection. The power supply voltage must be large enough to produce a suitable collector-emitter p.d. and thus provide one of the conditions mentioned above for putting the transistor into its operating region. In a typical radio receiver or tape player the supply voltage is between 3 V and 9 V, although some receivers operate on a supply of only 1.5 V.

The next consideration is the **quiescent state** of the amplifier. This is the state of the amplifier when it is operating but when it is not receiving any signal to amplify. In the quiescent state there is a constant base current and a constant (and larger) collector current. The collector is at a constant output voltage, measured with reference to the 0 V line. It is usual for the quiescent collector voltage to be half the supply voltage. When a signal is being amplified, the collector voltage rises and falls. If it is normally half-way between 0 V and the supply, it is able to rise and fall freely and equally in both directions. This makes it possible to have a large output signal without distortion. Usually, except in power amplifiers, the quiescent collector current is a few milliamps. Given this current, the value of R3 is selected to drop the voltage at the collector to the half-way level.

The transistor is **biased** into its operating region by the two **biasing resistors** R1 and R2. These act as a potential divider which provides sufficient base current to produce the required quiescent collector current. It is possible to bias a transistor with a single resistor either connected as R1, or between the collector and base, but biasing with two resistors gives the amplifier greater stability.

The signal is fed into the amplifier through the **coupling capacitor** C1 (p. 46). An alternating signal arrives at the input and passes across the capacitor. It causes a small and varying current to be added to or subtracted from the constant base current supplied by R1/R2. In some circumstances, it is possible to omit the capacitor and connect the amplifier to the signal source, but capacitor coupling is more often employed. For example, as we shall see in the two-stage amplifier, the quiescent output voltage (half the supply voltage) of the first stage would bias the transistor of the second stage into saturation if it were directly connected. The coupling transistor allows there to be a constant p.d. across the transistor without affecting the transmission of alternating signals across it.

Input impedance

The potential at the junction of the two resistors of a potential divider depends on their *ratio*. For example, it might be found that a suitable base current is provided by making R1 equal to 220 kΩ and R2 equal to 18 kΩ. If the supply voltage is 9 V, the potential at the base is $9 \times 18\,000/238\,000 = 0.68$ V. But the same potential could also be obtained if the resistors were 22 kΩ and 1.8 k, or even as small as 220 Ω and 18 Ω. In each case, the ratio of the resistances is the same.

Using smaller resistors would not affect the biasing of the transistor but it would have a serious effect on the input signal. Low-value resistors would short-circuit most of the tiny signal current to the 0 V or +V rails. Little of it would pass to the base. Most of the signal would be lost.

If the input resistance of the amplifier is too low, the signal is mostly lost. The same idea applies when we consider the fact that the capacitor, together with R1 and R2, make up a high-pass filter (p. 65). With certain values of the capacitor and resistors, the low-frequency portion of the signal is blocked and never reaches the transistor or, at least is appreciably reduced in amplitude. The combined resistive and capacitative effects on the input side are the **input impedance**. For the maximum signal to reach the transistor and be amplified, the input impedance must be as high as is feasible. This is achieved by making R1 and R2 as high as possible (if they are too high, the base current will not be large enough) and to select a value for the capacitor so that all required low frequencies are passed.

Output impedance

This is the impedance offered to the flow of current from the output side to the next stage of amplification or to a loudspeaker. In general, this should be as low as possible. The value of R3 should not be high, otherwise it may be impossible for enough current to flow from the supply line through R3 and on to the next stage. Similarly, if there is a coupling capacitor on the output side, the high-pass filter that it makes with R3 must pass all required low frequencies.

Input and output impedances vary with frequency. When specifying the performance of an amplifier, they are usually quoted for a signal frequency of 1 kHz.

Impedance matching

As explained above, if a circuit has low input impedance, a signal fed to it from another circuit is likely to be partially or almost wholly lost. Similarly, a circuit with high output impedance will not produce sufficient current to drive a subsequent stage of low input impedance. It can be shown that, for maximum transfer of power between two stages of a circuit, the output impedance of the first stage must equal the input impedance of the second stage.

Impedance matching is essential if power is not to be lost. One way of achieving this is to select suitable values of resistors and other components at the output and input. Most instances in which impedance matching is required involve feeding a low-impedance input from a high-impedance output. This can be done by linking the two circuits by an emitter follower amplifier (p. 99).

Two-stage amplifier

The amplifier below has two stages, connected one after the other **in cascade**. It does not simply amplify a signal and then amplify it still further. The function of the first stage, based on Q1, is to amplify the *voltage* of the signal produced by the microphone. The function of the second stage, based on Q2 is to amplify the *current*, making it sufficient to drive the loudspeaker.

The input from the crystal microphone is connected directly to the base of Q1. A crystal microphone has a high resistance so connecting it in parallel with R2 does not have any appreciable effect on the potential-divider action of R1 and R2. It does not upset the biasing of Q1 so there is no need for a coupling capacitor. Q1 is a high-gain transistor to amplify the input voltage from the microphone.

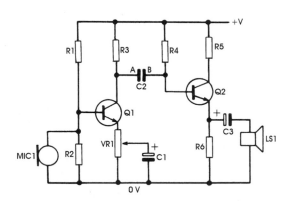

This two-stage amplifier comprises a voltage amplifier followed by a current amplifier.

With the collector of Q1 quiescent at around V/2, it is obvious that a capacitor C2 must be used to couple this stage to the next. The next stage has only a single bias resistor R4. R5 and R6 are chosen to bring the quiescent potential at the emitter of Q2 to the half-way point. Their values are relatively low to enable them to pass relatively high currents. Q2 is a low-gain power transistor able to pass a substantial current without over-heating.

The low values of R5 and R6 give the second stage a low output impedance, which means it can deliver a large current to the final stage, the loudspeaker.

With the collector of Q1 quiescent at around V/2, it is obvious that a capacitor C2 must be used to couple this stage to the next. The next stage has only a single bias resistor R4. R5 and R6 are chosen to bring the quiescent potential at the emitter of Q2 to the half-way point. Their values are relatively low to enable them to pass relatively high currents. Q2 is a low-gain power transistor able to pass a substantial current without over-heating.

The low values of R5 and R6 give the second stage a low output impedance, which means it can deliver a large current to the final stage, the loudspeaker. Loudspeakers typically have very low impedance, sometimes as low as 4 Ω and rarely higher than 100 Ω, so they need a large current to drive them. As the current through R6 fluctuates, a varying p.d. is developed across it. This produces a signal at the junction of R6 and the emitter and hence at C3. The signal passes across C3 and drives the loudspeaker.

The action of C1 is as follows. When the potential at the base of Q1 rises, base current increases and so does the collector current. The increased collector current passes through Q1 and VR1. Increased current through VR1 generates an increased p.d. between its ends. Its 'lower' end is fixed at 0 V, so the potential at its 'upper' end rises.

The factor that decides how much base current flows to Q1 is the base-emitter p.d. We have just explained that a rise in base potential results in a rise in emitter potential. As the base current rises and the base-emitter p.d. is *increased*, the emitter potential rises too and the base-emitter p.d. is *reduced*. The actions are in opposite directions. We call this kind of action **negative feedback**. It could happen that the rises in base potential and emitter potential were equal. In such an event the base-emitter p.d. remains constant, so the base current remains constant and likewise the collector current. The signal is completely damped out. No sound is heard.

The function of C1 is to absorb *part* of the change in base-emitter p.d., so that the amount of feedback can be controlled. The capacitor taps off part of VR1. Fluctuations in the p.d. across the tapped-off part are damped out by the action of the capacitor. Only the fluctuations in the part of VR1 between the emitter and the tapping are available as feedback.

By allowing a certain amount of feedback, the gain of the amplifier is reduced. This seems to be a fault, but it is compensated for by the fact that the fidelity of the amplifier is much improved. Its gain is constant over the whole of the frequency range and it is not affected by signal amplitude. In other words, it has a level response. In addition, the gain is not affected by temperature, as it is when there is no feedback.

FET amplifier

As explained on page 85, an FET has high input impedance. This makes it very suitable for use as the first stage of an amplifier that is to receive its input from a very high impedance source, such as a capacitor microphone (p.119). In the circuit below, the current through R2 causes a p.d. to develop across R2 when it is in the quiescent state. For reasons given earlier, the value of R2 is chosen so that the p.d. is about half the supply voltage. In other words, the source terminal of Q1 is at a positive potential. The gate is held at zero potential by R1. Consequently, the gate potential is negative with respect to the source potential, as required for the operation of the transistor.

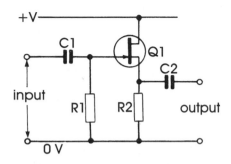

A JFET source-follower amplifier.

The signal passes through C1 to the gate, making the gate potential fluctuate and resulting in a varying current through the transistor. Note that although the gate of Q1 offers high input impedance, the impedance *of the amplifier* is the impedance between the input terminal and the 0 V rail. This impedance is the resistance of R1, typically 1 MΩ, which is high enough for most applications.

The transistor and R2 both have low resistance (a few hundred ohms) so the output of this amplifier has low impedance. Capacitor C2 couples this amplifier to later stages which may be based either on bipolar or field effect transistors.

This circuit is a common-drain or source-follower amplifier. It does not produce any voltage gain but is invaluable for connecting a high-impedance source with a low-impedance amplifier. High-impedance capacitance microphones of the **electret** type frequently have an FET amplifier built in to them, powered by a small dry cell. This matches the very high impedance of these microphones to the medium-impedance amplifier or tape recorder to which they may be connected. When such microphones are used at the ends of long screened cables, the high-impedance of the microphone acts in combination with the capacitance of the screened cable to produce, in effect, a low-pass filter. The filter reduces the treble component of the signal, giving a muffled sound after amplification. This type of distortion is reduced by the use of the FET pre-amplifier in the microphone case, for its medium-impedance output is not affected by cable capacitance to the same extent.

MOSFET amplifiers

A typical MOSFET common source amplifier is shown below. The transistor is an n-channel MOSFET so it operates with its gate at a positive potential. The bias is obtained from a potential divider as in the BJT amplifier on page 183. The gate requires virtually no current to bias it. R1 and R2 may each be several hundred of kilohms and a third resistor R3 may be placed between the potential divider R1/R2 and the gate. R3 may have a resistance as high as 10 MΩ, so the input impedance of the amplifier is in excess of 10 MΩ, which is very high indeed.

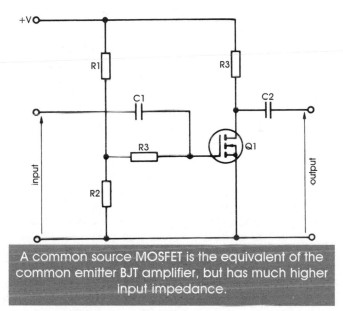

A common source MOSFET is the equivalent of the common emitter BJT amplifier, but has much higher input impedance.

Low-power MOSFETs have a rapid response time, making MOSFET amplifiers useful in radio-frequency circuits. Amplifiers based on power MOSFETs (such as VMOS and HEXFETs, p. 82) are recommended for power-control circuits. They have greater thermal stability than bipolar transistors and are not subject to thermal runaway (p. 70). They may be wired in parallel to enable very large currents to be controlled, in fact a HEXFET really consists of numerous paralleled MOSFETs.

Low-noise amplifiers

In the electronic sense, **noise** is an unwanted signal that is superimposed on a wanted signal. Much of a designer's effort goes toward reducing it.

Electronic noise may become apparent as actual noise, in the more usual sense, when we hear a crackling, humming, or hissing sounds in the background when listening to the radio. Crackling may be caused by electrical equipment in the neighbourhood switching on and off. Refrigerators are a common source of such noise. The spikes on the mains supply travel along the mains wiring and enter the power supply of the radio set. One way to reduce such **interference** is to pass the power supply leads through ferrite beads (p. 53).

Lightning is a powerful source of **electromagnetic interference** (EMI) in radio transmissions and little can be done to eliminate it. Any sort of magnetic field can be picked up by conductors in a circuit and, if it is picked up in a suitable place, can be amplified. Connections such as microphone cables are made with screened leads to prevent such EMI overwhelming the small signals from the microphone. The screen consists of a sheath of metallic braid, connected to the earthed (0 V) line of the amplifier. Tuning coils (p. 55) are usually cased in a metal shield, which serves to prevent the coil from either picking up EMI from close-by circuitry, or radiating it to other parts of the circuit. In today's congested conditions the subject of EMI is an important one and much legislation exists to limit emissions of EMI.

Precautions can usually reduce or eliminate noise coming from outside a circuit, but there are other sources of noise that arise within the circuit itself. An electric current is a stream of electrons or holes. There is a random element to their movements, giving rise to small random fluctuations in *voltage*. When amplified, this can produce a hissing sound, superimposed on the normal signal. This is often known as **white noise**. This type of noise is produced whenever current flows through any kind of resistance. To minimize it, we must try to reduce the resistance of circuits, but this is not always practicable. The random movements are greater at higher temperatures, so keeping an amplifier cool will help reduce noise. Another source of noise is that even with a steady current the actual number of particles moving past a given point in a given time is not absolutely constant. There are random fluctuations in *current* and these too lead to noise.

In non-audio circuits, noise can show itself in appropriate ways, for example as 'snow' on a TV screen. In a control system, it may be manifest as erratic responses. In a security system, it may lead to false alarms. In a digital system, noise may result in a digital '0' being read as a '1', and the other way about, with possibly disastrous results.

Noise can be added to a system at any stage but the biggest problems arise when it is added at an early stage. If the early stage of an amplifier is noisy, the noise is amplified along with the signal and may be very difficult to eliminate.

It is therefore very important to concentrate on the early stages of amplification. Low-noise transistors, reduction of resistance as much as possible, and filtering of the signal to remove all except the desired signal are all measures that can be taken to reduce noise. FETs produce less noise than BJTs and so are often used in the early stages of amplification, then may be followed by BJTs.

A good example of the need for low-noise amplifiers is illustrated in the title photograph of this chapter. This is one of the smaller radio-telescopes at Jodrell Bank. The signal reaching the telescope from Space is extremely weak and all possible noise-avoidance techniques are used. The antenna and the first stages of amplification are located at the focal point of the reflector, mounted on a structure of four girders. The signal is amplified immediately it is received, before it can be contaminated with signals from terrestrial sources. Various amplifiers are used at the focus, all mounted on a carousel inside casing. Amplifiers can be brought to the focal point one at a time on a rotating head. These are narrow-band amplifiers, tuned to operate at a single frequency. This is a good way of limiting noise since some types of noise are proportional to band-width. A narrow-band amplifier is much less susceptible to noise than a wide-band amplifier. The amplifier itself may be a special type known as a **parametric amplifier**, which has low noise properties. Another noise-reduction technique employed at Jodrell Bank is to cool the receiver and amplifier to 14 K (−259°C) with liquid helium. Low temperature reduces the random vibrations of the electrons, so reducing a significant noise source.

18 Operational amplifiers

The original operational amplifiers were built from discrete components but, since the development of integration, all operational amplifiers (or, to give them their more usual name, **op amps**) are in the form of integrated circuits. There are dozens of different types, some for general use, others intended for specific purposes. One of the most popular general purpose op amps is the '741', featured above. All operate according to the same principles.

Op amps are not the only type of integrated amplifier obtainable. There is a whole range of IC amplifiers manufactured for special purposes, including audio amplifiers (mono and stereo, pre-amplifiers and power amplifiers), VHF amplifiers, UHF amplifiers, video amplifiers, and pre-amplifiers for infra-red sensors. Each of these has its own special features of concern to the person who is using them, but not of general interest. Op amps and their features are of general interest and such wide application, that they deserve a whole chapter to themselves.

The features of an op amp are as follows. They all have two inputs, known as the **inverting input** (–) and the **non-inverting input** (+). They operate relative to the 0V line but they usually take their power supply from a positive line (say, +9 V) *and* a negative line of equal but opposite potential (–9 V). There is a single output terminal.

The connections to the power lines are omitted in the figures in this chapter as are connections to certain terminals such as the offset null pins. These are provided on the IC to allow the output of the amplifier to be set to zero when both inputs are at the same potential. In other words, to allow for tolerances in manufacture. With many present-day amplifiers offset errors are extremely small and the terminals for adjustments are not provided.

One important point about op amps is that the inputs have very high input impedance. Those with bipolar transistors at the inputs have an input impedance of about 2 MΩ. Many types of op amp have FET or MOSFET input stages. For the latter, the input impedance is 10^{12} Ω, or a million megohms. This is a virtually infinite input impedance. In contrast, the output impedance of op amps is usually very low, of the order of 75 Ω.

The gain of an op amp is very high, typically 100 000. This is known as the **open loop gain**. Although we sometimes make use of this high gain, the op amp is more often connected so that the gain *of the circuit* (as opposed to the gain of the op amp itself), the **closed loop gain**, is appreciably less than this. We return to this point later.

To shorten the explanations in this chapter, we will refer to the non-inverting input as the (+) input and to the inverting input as the (−) input. These are also the symbols used to identify these inputs on the circuit diagrams. All potentials are measured with respect to the 0 V line. Remember that all potentials applied at the inputs and all output potentials can never be greater than the positive supply voltage or less than the negative supply voltage. In some types of op amp, output voltages have a more restricted range and can not lie within a volt or two of the supply voltages.

An op amp is a **differential amplifier**. Its output depends on the *difference* between the potentials at the two inputs. If the (+) input is at a higher potential than the (−) input, the output swings positive. If the (−) input is at a higher potential than the (+) input, the output swings negative. If the two inputs are exactly equal, the output stays where it is. Because the gain is so high, the difference between input potentials is always exceedingly small, almost zero. To put it the other way round, if the difference is more than a few millivolts, the amplifier saturates and the output swings as far as it will go positive or negative.

Note that in the remainder of the chapter we use the term *op amp* to mean an IC, such as a 741. We use the term *amplifier* to mean an amplifying circuit built from one or more op amps, plus various resistors and possibly other components.

Using op amps

Op amps are handy building blocks used in a wide range of circuits. The figure below shows an op amp being used to build an **inverting amplifier** circuit. It has two resistors, the input resistor R1, and the feedback resistor R2. The (+) input is connected to the 0 V line.

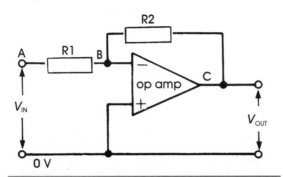

In an inverting amplifier, there is feedback from the output of the op amp to its inverting input.

If a positive voltage V_{IN} is applied to the (−) input through terminal A and R1, the output V_{OUT} swings negative. Point A is positive and point C is negative so somewhere in between them there must be a point which is at 0 V. The output will stabilize if the 0 V point is at B, for then the inputs of the op amp are at equal potential (0 V).

Under these conditions a current I flows from A, through R1 to B. We have already said that this has extremely high input impedance so virtually no current flows into B. Therefore, I continues unchanged through R2 to the output C. Along R1 the potential drop is V_{IN} and:

$$I = \frac{V_{IN}}{R1}$$

Along R2 the potential drop is V_{OUT} and:

$$I = \frac{-V_{OUT}}{R2}$$

The negative sign is needed because V_{OUT} is negative. Since the current I through R1 is the same current as passes through R2, and since the term I refers to the same current in both equations, we can put the right-hand sides equal to each other:

$$\frac{V_{IN}}{R1} = \frac{-V_{OUT}}{R2}$$

Rearranging the terms of this equation, we obtain:

$$V_{OUT} = -V_{IN} \times \frac{R2}{R1}$$

The final equation shows that the gain of the circuit is $-R2/R1$. The negative sign indicates that if the input is positive, the output is negative, and the converse. In other words, this is an inverting amplifier. An important fact is that the gain depends only on the values of the two resistors, and not on the open loop gain of the actual op amp used. For example, if we make $R2 = 2.2$ MΩ, and $R1 = 10$ kΩ, the gain is $2\,200\,000/10\,000 = 220$. By choosing suitable resistor values we can obtain any required gain that is not as great as the open loop gain of the amplifier.

In the inverting amplifier, the current flowing along R1 is entirely diverted along R2. Although the inverting input has high impedance it is always automatically brought to 0 V by appropriate voltage changes at the output. The input behaves as if it were an 'earth', that is to say, any current arriving at the input 'disappears' (actually, along R2) just as though point B was connected directly to earth. We call the input terminal a **virtual earth** in this circuit. Since current entering terminal A is, in effect, directed to earth at the virtual earth, the input impedance of this circuit equals R1.

We make more use of the virtual earth in the next circuit, which is an **adder**. The four (in this case) quantities to be added are represented by different potentials, V_1, V_2, V_3, and V_4. If these are applied to the four inputs of the adder circuit, currents proportional to these potentials flow along the equal input resistors.

No current can enter or leave the inverting input so the combined currents all flow along the feedback resistor (also equal to R). To bring the potential at B to zero the output must fall by an amount equal to the sum of the four input potentials. We can write the result as an equation:

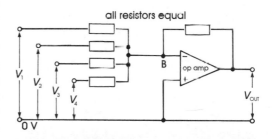

$$V_{OUT} = -(V_1 + V_2 + V_3 + V_4)$$

An extension of the principle of the inverting amplifier gives a circuit that adds several quantities. Its output is the inverse of their sum.

This principle may be used in a circuit in which the output is to be obtained by summing two or more values measured by sensors. It is also used in audio equipment as a mixer, summing signals from two or more sources, such as two tape players. By using variable resistors instead of fixed resistors, it is possible to weight the sum, in effect multiplying the signals by different factors to mix in more or less of each signal.

The third example of an op amp circuit is a **non-inverting amplifier**, with the input connected through R1 to the (+) input of the op amp. R1 is needed in high-precision circuits but generally it makes little difference if R1 is omitted. Only a minute current flows into the high-impedance input and therefore there is a negligible voltage drop across it.

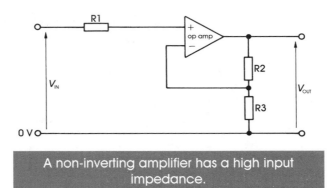

A non-inverting amplifier has a high input impedance.

With a given potential V_{IN} at the non-inverting input, the amplifier output stabilizes when this same potential is present at the inverting input. This is the same potential as at the junction of R2 and R3. These two resistors may be considered as a potential divider with V_{OUT} applied across it and producing a potential equal to V_{IN}. From the equation on page 60:

$$V_{IN} = V_{OUT} \times \frac{R3}{R2 + R3}$$

Rearranging the terms of this equation gives:

$$V_{OUT} = V_{IN} \times \frac{(R2 + R3)}{R3}$$

For example, if R2 = 220 kΩ and R3 = 10 kΩ, the gain is (220 000 + 10 000)/10 000 = 23. Note that the gain is *positive* as this is a non-inverting amplifier. This configuration of the op amp is used when inversion of the signal is undesirable.

A particular advantage of the circuit is that it has a very high input impedance, equal to the impedance of the op amp input itself. For a really high input impedance we may use an op amp with JFET or MOSFET inputs, with an impedance of around 10^{12} Ω.

In a variation of the non-inverting amplifier the input to the circuit is connected directly to the non-inverting input. The output is connected directly to the inverting input. We can consider this circuit to be a non-inverting amplifier in which R2 is zero and R3 is infinite. As R3 approaches infinity, the value of the expression (R2+R3)/R3 approaches 1, or unity. Thus, the equation for this circuit is:

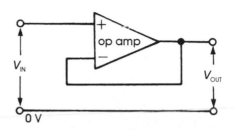

$$V_{OUT} = V_{IN}$$

A unity gain voltage follower is a simple way to match a high-impedance output to a low-impedance input.

Whatever potential is applied as V_{IN} (within limits) the same potential appears as V_{OUT}. We call this circuit a **unity gain voltage follower.** This circuit is excellent for impedance matching (p.185). Its input impedance is that of the op amp input, which is at least 2 MΩ, and more if the op amp has FET inputs. As a further advantage, the output impedance of the op amp is very low. This circuit has much the same features as the emitter follower (common collector amplifier, p. 99) though its performance is better.

Instrumentation amplifiers

An instrumentation amplifier (or **in amp**) can be built from three op amps connected as in the circuit overleaf, but it is more convenient and reliable to use an in amp consisting of three op amps ready connected on a single chip. Like an op amp, an in amp is a differential amplifier which amplifies the potential difference between its inputs. The inputs go directly to the inputs of two of the constituent op amps, so *both* inputs are of high impedance and are equal in value, typically 10^9 to 10^{12} ohms. This is not the case in a differential amplifier built from a single op amp, where the impedances are unequal and relatively low.

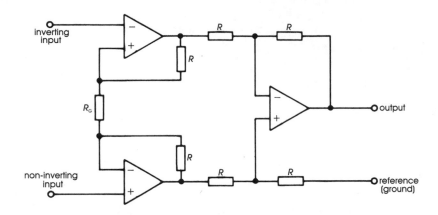

An instrumentation amplifier consists of three op amps; all resistors except R_G are equal.

Another advantage of the in amp is that all except possibly one of the resistors are fabricated on the same chip, so that they are more likely to be accurately balanced.

R_G may be an external resistor connected across two pins of the IC. If the pins are unconnected, the gain of the amplifier is 1 but different gains may be obtained by connecting a resistor of suitable value at this point. Alternatively, a number of resistors may be provided on the chip to give gains of 1, 10, 100 and 1000; one of these can be selected by connecting appropriate pins together.

In amps are used where long-term stability and high sensitivity are essential. Applications include medical instrumentation such as ECG and EEG circuits.

Transconductance amplifiers

A transconductance amplifier is similar to an op amp, having the same pair of inverting and non-inverting inputs, and the same positive and negative power supplies. It is in fact a type of op amp but with one important distinction, it is the output *current* (not the output voltage) that is controlled. The output current is proportional to the potential difference between the input terminals. In addition there is a bias input which may be used to switch the output current on or off, or to vary the transconductance of the amplifier.

Transconductance

The voltage gain of an amplifier is the ratio V_{OUT}/V_{IN}. Similarly, the current gain of an amplifier is the ratio I_{OUT}/I_{IN}. Both of these are ratios, so they are pure numbers with no units. For a transconductance amplifier the equivalent to gain is I_{OUT}/V_{IN}. This is obtained by dividing a current by a voltage, so it is not a ratio. When we calculate resistance we divide volts by amps (p. 25). In calculating the 'gain' of a transconductance amplifier we divide amps by volts. This is the inverse operation and has inverse units. Instead of *resistance* with *ohms* as the unit, we have *conductance* with *siemens* as the unit. Because the volts are at the input side of the amplifier and the current is at the output side, this particular type of conductance is known as *trans*conductance.

The transconductance amplifier is often used as an attenuator to control the amplitude of a signal. The signal is fed into the non-inverting input and, according to the potential supplied to the amplifier bias input, the amount of current flowing from the output can be adjusted. This may be fed directly to the base of a transistor for current amplification. A similar application is as an envelope shaper in electronic musical instruments. A piano, for example has a characteristic percussive sound. The note begins with high volume then dies away slowly. Its amplitude envelope shows a sharp rise (attack phase) and a slow fall (decay phase). A note of constant amplitude may be passed from a signal generator, through a transconductance amplifier, to have its envelope shaped to imitate that of a piano. By contrast, the note from a violin usually has a slow attack phase and often an equally slow decay phase. Envelope shaping adds greatly to the realism of electronically generated instrument sounds.

Amplifier gain

When engineers refer to the gain of an amplifier they generally express it in decibels (symbol dB). As explained in the box overleaf, gain is a pure ratio with no units. The decibel is not a unit of gain but a *scale* on which the gain ratio may be expressed. Given two quantities such as V_{OUT} and V_{IN}, the gain is the ratio between them, V_{OUT}/V_{IN}. In decibels, this ratio is expressed by taking its logarithm (to base 10) and multiplying by 10.

For example, if V_{OUT} = 4 V and V_{IN} = 0.5 V, the gain is 4/0.5 = 8. Expressed in decibels, this is:

$$10 \times \log_{10} 8 = 9 \text{ dB}$$

Because the decibel scale is a logarithmic one, doubling the gain does not double the decibel value. A gain of 16, for example, is equivalent to 12 dB, not 18 dB. Multiplying gain by 100, to make it 800, increases the decibel rating only to 29 dB. Although this scale is difficult to comprehend at first, it is a useful one for the audio designer. It is used not only for expressing gain but the ratio between any two quantities that are measured in the same unit. For example, we may express the ratio between mean signal level and mean noise level in terms of the ratio of their amplitudes and then convert this to decibels. It has the advantage that in a multi-stage circuit the decibel gains of each stage may be added together instead of being multiplied. For example, if an amplifier with a gain of 50 dB is cascaded with an amplifier with a gain of 100 dB, the total gain is 150 dB. The box on page 203 illustrates the usefulness of the decibel scale.

Power ratios

Ohm's Law tells us that $R = V/I$. Rearranging this equation gives $I = V/R$. The power being dissipated in a resistance is defined as $P = IV$. Combining the equations gives $P = V/R \times V = V^2/R$. Power is proportional to voltage squared. When we are more concerned with the power of the input and output signals of an amplifier rather than the voltage, we take the logarithm of the ratio of the mean input and output voltages and multiply by 20, instead of by 10. Doubling a logarithm has the effect of squaring, so this is equivalent to squaring the voltages *before* finding their ratio. It gives the ratio between output and input *power*, in decibels.

Active filters

The filters described on pages 62–5 are known as passive filters because they are built from passive components such as resistors, capacitors and inductors. An op amp is an active device, so filters based on op amps are known as active filters.

A simple example is shown below. Its active part is an op amp connected as a non-inverting amplifier (p. 196). By comparison with the filter on page 62, it can be seen that this filter has two low-pass passive filters in cascade.

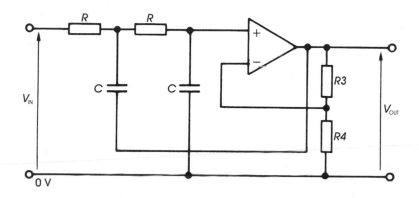

The op amp is the active element in this low-pass active filter.

The two resistors marked R are equal in value and so are the two capacitors. The signal thus passes through two low-pass filters on its way to the op amp, which is why this is called a **second-order filter**. The op amp amplifies the signal that reaches it; this is an improvement since, in a series of cascaded passive filters, the output always has a much lower amplitude than the input. But this is not the only benefit. The second filter has its capacitor grounded on the 0 V line, but the first stage has its filter connected to the output of the op amp. The output signal is being fed back to an earlier stage of the circuit.

It is a property of a single-stage passive filter that, at the cut-off frequency, the output signal lags an eighth of a cycle behind the input signal. Two stages of filtering give two eighths of a cycle of delay, or a quarter of a cycle. This delayed signal is fed back to the input stage and acts to reduce the amplitude.

At increasing frequencies the delay becomes greater and greater and the resulting output is more and more reduced. At very high frequencies the delay reaches a half-cycle. The input and output are out of phase. This means that when the input signal is going positive the output signal is going negative. When the input signal is going negative, the output is going positive. As a result of this half-cycle delay, the fed-back signal tends to damp out the incoming signal, thus reducing its amplitude to a very low level. High-frequency signals are almost completely eliminated.

The action of the filter is complicated but it can be illustrated by a graph on which we plot amplitude against frequency. We use a sinewave oscillator to generate a signal of constant amplitude. This is passed through a filter and we measure the amplitude of its output for a number of different frequencies. Here the frequency ranges from 10 Hz to 100 kHz. The frequency is plotted on a logarithmic scale, on which equally spaced graduations correspond to a ten-fold increase in frequency. This allows a wide frequency range to be shown on a single graph. The amplitude in volts is plotted in decibels, with respect to the input signal. This also allows a wide range of amplitudes to be plotted. Note that the amplitudes have negative values in decibels because they are all smaller than the input signal.

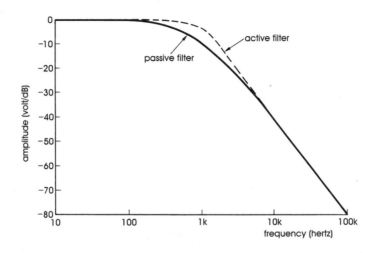

The frequency response of an active filter has a much sharper 'knee' at the cut-off frequency than a passive filter does.

To start with, study the continuous line, which shows the response of a second-order passive filter. The values of the capacitors and resistors have been selected to give a cut-off frequency of 1 kHz. The output at low frequencies is 0 dB, that is to say, it has the *same* amplitude as the input signal. Above 100 Hz, the output amplitude begins to fall. It falls to about −10 dB at the cut-off frequency, then continues falling until it eventually falls steeply at a steady rate. Measurements on the graph show that the rate of fall is −12 dB per octave. This means a fall of 12 dB for every doubling of frequency.

Using decibels

The frequency response graph is a good example of the way in which decibels are often a more convenient way of expressing ratios. The input signal to both filters has an amplitude of 1 V. The output of the passive filter has an amplitude of 1 V for low frequencies so, for the passive filter, 0 dB corresponds to 1 V. But the active filter has slight gain. At low frequencies its output is 1.59 V. For the active filter, 0 dB corresponds to 1.59 V. It is difficult to compare the two output curves if we plot them on an absolute voltage scale as the active filter output begins with a higher value. Plotting them on a decibel scale makes it easy to see precisely how the output responses differ.

The output of the active filter, with the same resistors and capacitors but with the addition of an op amp, plus R3 and R4, is shown by the dashed line. This curve has a much sharper bend in it than the passive curve. The filter passes frequencies just below the cut-off frequency much more readily than does the passive filter. The curve reaches its maximum rate of fall (12 dB per octave) much sooner. In other words the graph shows that the passive filter has a much narrower zone between those low frequencies that it passes readily and the higher frequencies that it attenuates. It is a much more effective filter. At the cut-off frequency, the signal is reduced to −3 dB, which corresponds to a reduction in signal power of exactly 50%.

The op amp has sharpened the response of the filter. Another way to do this is to include an inductor in the filter circuit, but the disadvantage of this is that inductors of suitable value may be both large and weighty. Also, unless carefully shielded, inductors tend to disseminate and pick up electromagnetic interference. This makes them unsuitable for use in modern compact and portable equipment. Op amps have almost completely replaced inductors in filter circuits.

19 Logic circuits

W hether we are reading the time on a digital watch, dialling a call on a mobile phone, listening to a compact disc, using a Smartcard, or sending an e-mail, we are using logic circuits. More and more of the equipment we use at home and at work depend on electronic logic. Logic is the science of reasoning, and we use its electronic equivalent as the basis for a whole host of so-called 'intelligent' devices.

The form of logic most suitable for implementing electronically is **binary logic**. In binary logic, we deal with only two states. A statement is *true* or it is *untrue*: there are no half-truths. A digit is 1 or 0: there are no fractions or other values. A transistor is either fully on (saturated) or fully off: it switches rapidly from one state to the other, spending a negligible time in the intermediate states. A voltage is either high or low: intermediate values have no meaning.

The on-off, high-low characteristics of binary logic mean that we can design electronic circuits that can model logical statements with absolute accuracy. We need to be concerned with only two voltage levels, not with all the possible levels that exist in (for example) audio circuits. Nowadays, even audio circuits are mainly digital but, at least, in their input and output sections, we have a continuous range of voltage levels.

Let us see how a logical situation may be modelled electronically. Take as an example the floodlight illustrated in the title photograph of Chapter 12. A floodlamp beside a garage is to be switched on whenever a car or a person approaches the garage, but this is to happen only at night. There is no point in turning on the floodlight during the day. There are two sensors. One of these, perhaps based on a light-dependent resistor responds to light level. It can be arranged, possibly by using a Schmitt trigger circuit, that the output voltage of the sensor circuit is low (close to 0 V) when it is daylight and high (close to the supply voltage) at night. The other sensor is a pyroelectric sensor; it responds to the heat radiating from the car or from the body of the approaching person. Its circuit is arranged to give a high output when it is triggered.

The logic required is simple. If the LDR detects night AND the pyroelectric device detects heat, the floodlight comes on. The AND in this sentence is an example of a **logical operator**. It links the two conditions under which the floodlamp is to be turned on. It tells us that we must have one condition (night) AND the other condition (heat detected) in order to turn on the floodlamp. Putting it another way, we have three statements:

A = it is night
B = a car is approaching
Z = the floodlamp is on

Any of these statements may be true or not true, but the truth or otherwise of Z depends on the truth of A and B. The logic of the lighting control, expressed in short form is:

IF A is true AND B is true THEN Z is true

or even shorter:

IF A AND B THEN Z

The 'IF ... THEN ...' format appears often in binary logic.

Inverse statements

Each of the statements above can be paired with an 'opposite' statement so that, if the statement is not true, its 'opposite' or inverse is true. The inverse is represented by the same letter as the original statement, but with a bar over it. For example, if A= 'it is night', then $\bar{A}$ = 'it is not night'.

In most contexts, including the example above, we accept that 'it is not night' means the same as 'it is day'. But, at dawn and dusk, people could be justified in stating that it is not night, yet it is not day either. So, 'it is day' is not a strict inverse of 'it is night'. The only exact inverse is 'it is not night'. This difficulty does not arise with Z = 'the floodlamp is on' , which has the inverses $\overline{Z}$ = 'the floodlamp is not on', and $\overline{Z}$ = 'the floodlamp is off' because 'off' and 'on' are exclusive opposites. Because we have additional knowledge about floodlamp systems, we are able to use the other inverse. However, if in doubt, always make the inverse by inserting 'not'.

Logical operators

There are three basic logical operators that are used in 'IF ... THEN ... ' statements:

NOT: IF A is true, THEN $\overline{A}$ is NOT true. Since all statements must be true or not true, this implies that if A is NOT true THEN $\overline{A}$ is true.

Example IF it is night THEN it is NOT daytime (assuming we ignore dawn and dusk).

AND: IF A is true AND B is true, THEN Z is true. Otherwise Z is not true (false).

Example IF it is night AND a car approaches, THEN the floodlamp comes on.

OR: IF A is true OR B is true, THEN Z is true. Otherwise Z is not true (false).

Example At night, if a person approaches OR a car approaches, THEN the floodlamp comes on.

The full operation of the floodlamp controller depends on the AND and OR operators. We shall shortly look at ways in which these operations can be performed electronically.

Truth tables

The action of logical operators can be represented by truth tables in which '1' represents 'true' and '0' represents 'not true', or 'false'. Alternatively, we could use the letters 'T' and 'F'. We list the 'IF' conditions A, B under 'Input' and the 'THEN' consequence Z under 'Output'. On the opposite page are the three tables that cover the logical operators we have just discussed.

NOT	
Input A	Output Z
0	1
1	0

AND		
Input A	Input B	Output Z
0	0	0
0	1	0
1	0	0
1	1	1

OR		
Input A	Input B	Output Z
0	0	0
0	1	1
1	0	1
1	1	1

Truth tables of the three basic logical operators.

Except for the NOT operator, logical statements can be extended to include more than two conditions. For example, IF A is true AND B is true AND C is true AND D is true, THEN Z is true. Otherwise, Z is false.

It is a simple matter to perform the basic operations using ordinary manually operated switches, with a lamp to indicate the logical outcome. This is how we perform the AND operation on A, B, C and D.

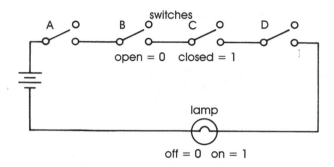

In the AND operation, the lamp is on only if all four switches are closed.

For calculators, clocks and computers we need full electronic switching at high speed, combined with reliable action, freedom from interference from occasional voltage fluctuations and 'spikes', and outputs that are firmly high or low. Logic circuits that meet these specifications are made in integrated form and we consider some of these in the next section.

All logical situations can be expressed in terms of NOT, AND and OR, but there are three other operators that are often used to simplify the statements:

EXCLUSIVE-OR, sometimes known as **EX-OR**: If A OR B but NOT BOTH are true, then Z is true.

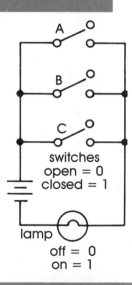

In the OR operation, the lamp is on when any one or more switches is closed.

NAND: This is the AND operation followed by NOT.
NOR: This is OR followed by NOT.

Exclusive-OR		
Input A	Input B	Output Z
0	0	0
0	1	1
1	0	1
1	1	0

NAND		
Input A	Input B	Output Z
0	0	1
0	1	1
1	0	1
1	1	0

NOR		
Input A	Input B	Output Z
0	0	1
0	1	0
1	0	0
1	1	0

Three often used logical operators.

All logical operations can be performed using just the NAND operator or the NOR operator. Statements using only NAND or NOR may be longer and more difficult to understand that those which use the simpler AND, OR, and NOT operators, but there is sometimes a practical advantage. It can be more economical when building a logical circuit to employ only one type of operator and the wiring may be simpler.

Transistor-transistor logic

Although transistor-transistor logic (or **TTL**, as it is more usually called) has been largely superseded by other forms with superior performance, it is TTL which was the dominant type in the expanding days of the seventies and early eighties. Many of the logic ICs today are simply upgraded versions of the original TTL, performing the same functions and having the same pin connections.

The circuit element that performs a single logical operation is known as a **gate**. The circuit has an unfamiliar component which is a transistor with two emitters. If both emitters are connected to the positive supply voltage (logical high, equivalent to A AND B) the transistor does not conduct. The effect of this on the remainder of the circuit is to make the output Z go low. The transistor is turned on when either one or both emitters are connected to 0 V (logical low), any combination of inputs making the output go high

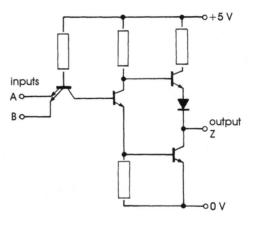

The circuit of a 2-input NAND gate in transistor-transistor logic.

The behaviour of the circuit mirrors the action of the NAND operator. One might wonder why it requires such a complicated circuit to perform this operation. One function of the other components is to ensure fast action. The other function is to ensure that the voltages at high and low levels are clearly distinguished, with no half-way states possible.

Even though the circuits of the descendants of TTL are appreciably more complicated, they are still simple enough from the point of view of integration, and they occupy little space on the silicon chip. Four independent gates are fabricated on one chip in a 14-pin package.

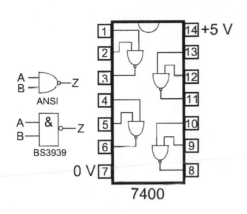

7400

Four separate NAND gates are contained in a single 7400 IC, which has 14 pins. The American and British symbols for a 2-input NAND gate are shown on the left.

TTL and its derivatives comprises a large family of several hundred different ICs. Some of these consist of relatively simple gates but others employ MSI and LSI (p. 164) to build complex counters, registers and logical arrays on a single chip. The chief advantage of TTL is its high speed, making it suitable for use in computers. It requires a supply voltage between 4.75 V and 5.25 V so a regulated supply is virtually essential. Electrical noise such as 'spikes' on the supply lines are liable to upset its operation. The power supply needs to be well filtered to remove this interference.

Its main disadvantage is that it requires fairly large currents to operate it. Even a relatively small system consisting of half-a-dozen ICs may require a supply of 1 A, which means that the use of TTL in battery-powered equipment is severely limited. This has prompted the development of versions of TTL that are less power-hungry. One of the most popular of these is low-power Schottky TTL, or LSTTL. Power requirements are reduced to about a fifth of those of standard TTL by increasing the values of the resistors in the circuits. The name 'Schottky' refers to an element in the circuit known as a Schottky clamp. The function of this is to prevent transistors that are turned on from being fully saturated. Saturated transistors take longer to switch off than transistors that are not saturated because it is first necessary to remove excess charge carriers from the base region of the transistor. The Schottky clamp forward-biases the transistor by only a few tenths of a volt so that it turns off quickly. The LS series is not much faster than standard TTL, but the newer Advanced low-power Schottky version is twice as fast and uses about one-twentieth of the power of standard TTL.

CMOS logic

Although standard CMOS logic (the best known is the 4000 series) operates about 10 times more slowly than TTL and its variants, this is no disadvantage in many applications. Its chief advantage is that it has very low power consumption, making it ideal for battery-powered portable equipment. When gates are not actually changing state they require practically no current at all, so quiescent power requirements are very low indeed. CMOS operates on any supply voltage between 3 V and 15 V. A regulated power supply is not required. An LSTTL output can drive only about five LSTTL inputs and similar limitations apply to the other TTL versions. By contrast, a CMOS output can drive a practically unlimited number of CMOS inputs. This makes circuit design much simpler.

The 'C' in the acronym CMOS stands for 'complementary' and the figure explains what this means. The gate consists of two MOSFETs of complementary type, n-channel and p-channel. Given a high input, the p-channel transistor is turned off and the n-channel input is turned on. This more or less short-circuits the output to the 0 V rail, giving a low output. Conversely, a low input turns on the p-channel transistor and turns off the n-channel transistor, short-circuiting the output to the positive supply rail. The action corresponds to the logical NOT operator.

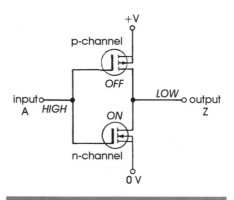

A high input connects the output to the 0 V line.

The NAND gate illustrated on the opposite page operates in a similar fashion. The output is connected to the 0 V line only if both inputs are high, switching on both n-channel transistors. Other combinations of inputs connect the output to the positive supply line.

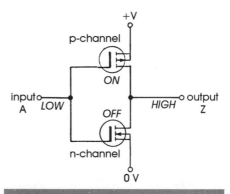

A low input connects the output to the positive supply line.

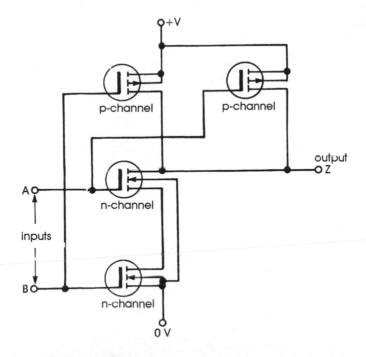

Four 2-input CMOS NAND gates like this one are contained in the 4011 IC.

Because CMOS works by short-circuiting the output terminal to one or other of the power rails, the output voltage is very close to one or other of the supply rails. There is a clear distinction between the high output level and the low level. This is why CMOS is relatively unaffected by noise and 'spikes' on its supply lines.

The diagram also illustrates the fact that the CMOS gate requires few components compared with a TTL gate. Thousands of gates can be accommodated on a single chip so that CMOS is ideal for LSI and VLSI. Consequently the CMOS range of logic circuits includes devices such as 16-stage counters and 64-stage shift registers that are not available in TTL.

CMOS transistors, like all MOS transistors, have very high input impedance, so they require virtually no current to alter the potential of the gate. We can understand why the current requirements of CMOS are exceedingly small, and why current is required only when the input to a gate is being changed.

Transmission gates

In this type of gate, the n-channel and p-channel transistors are both switched on by a high control input and are both switched off by a low control input.

The circuit acts as an electronically-controlled switch and can be controlled by the output from another CMOS gate. When the transistors are off, only a minute current of a few nanoamps can flow between the terminal on the left to the terminal on the right. When the transistors are on, the resistance between input and output terminals is only about 80 Ω, the gate acting as a low-value resistor.

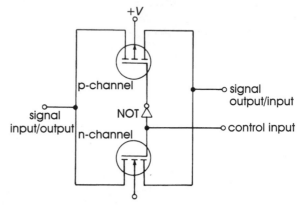

Transmission gates are used for switching analogue signals under digital control.

When the gate is switched on, current can flow freely through it in either direction, which is why we refer to its terminals as *input/output* terminals. Transmission gates are used for switching analogue signals (for example, audio signals) may range in size between 0 V and the supply voltage.

The gate shown above acts as a single-pole on-off switch. Other CMOS transmission gate ICs have two or more transmission gates with common control lines. These make it possible to implement other switching actions, such as changeover switches, double-pole double-throw switches, and 1-of-4 (or more) selector switches, all under logical control.

Other logic

Among the other logic families, each with its own characteristics and special applications, is high-threshold logic (HTL) or high noise immunity logic (HNIL). The gap between the low and high logic levels is particularly large, which makes the series suited to industrial environments. There, electrical noise often makes it impractical to use TTL or CMOS.

Emitter-coupled logic (ECL) is designed for very high-speed operation; as with Schottky devices, the transistors are prevented from becoming completely saturated, but in this family this is achieved by having the logical low and high levels close together. This makes ECL liable to noise interference. Also, it requires more power, as the transistors are bipolar types and are always switched on. The HC family is very popular. It has CMOS-based gates but runs at high speed. It combines the best features of TTL and CMOS. Most of the standard TTL devices are available as HC types, and there are HCT devices intended for interfacing directly with TTL.

Logic and numbers

Electronic logic involves only two states; on and off, high and low, 1 and 0. But the numbers we use in everyday life rely on having 10 different symbols (0 to 9) to express them. To convert these directly to electronic form would mean having 10 input and output levels. Although it is *possible* to do this, the circuitry required is much more complicated than that used for logic gates. We throw away the advantages of high-speed operation, reliability, noise-immunity, and the ease of implementing the circuits in MSI and LSI. It is much more sense to abandon the decimal system and replace it with a number system that has only two values, 0 and 1. This is known as the **binary system**.

Decimal number	Binary equivalent
0	0
1	1
2	10
3	11
4	100
5	101
6	110
7	111
8	1000
9	1001
10	1010

In the decimal system, the value of a digit is determined by its place in the number. For example, the number '374' means 3 hundreds *plus* 7 tens *plus* 4 units. In the number '6267', the first '6' means 6 thousands, but the second '6' means 6 tens. Value depends on position. As we go from right to left, the position value increases by 10 times.

The binary system works on the same principle but in twos instead of tens. The position values from right to left are units, twos, fours, eights, sixteens and so on, doubling up each time we shift one place to the left. So the binary number 110 is the equivalent in decimal of 1 four *plus* 1 two *plus* no units, totalling $4 + 2 + 0 = 6$. Similarly, the binary number 1010 is the equivalent of 1 eight *plus* no fours *plus* 1 two *plus* no units $= 8 + 0 + 2 + 0 = 10$.

Bits and bytes

The short word for a binary digit is bit. Thus, the number 101 has three bits. The bit on the right, the unit bit, is known as the least significant bit. The bit on the left, the 'fours' bit in this example, is known as the most significant bit. A string of eight bits is known as a byte.

Mathematics is a logical process so it is not surprising that we are able to use logic circuits to perform mathematical operations. As a simple illustration, consider the addition of two single-digit binary numbers, either of which may be 0 or 1. The results of all possible additions with these two numbers are:

INPUT		OUTPUT	
A	B	sum, S	carry, C
0	0	0	0
0	1	1	0
1	0	1	0
1	1	0	1

The addition in the last line of the table would normally be written with the carry digit on the left: 1 + 1 = 10. In decimal, this is the equivalent of 2 + 2 = 4 as can be seen from the table on page 215.

If we compare the value S in the sum column with the output values of the truth tables on page 188 we find that S is the equivalent of Z for an EXCLUSIVE-OR gate.

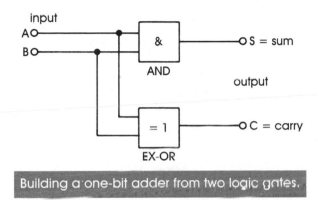

Building a one-bit adder from two logic gates.

Similarly, matching the value of C in the carry column, we find that this is the same as Z for an AND gate. Given an EXCLUSIVE-OR gate and an AND gate, we can use them to add two single-digit binary numbers. Addition of binary numbers with more than a single digit is naturally more complicated, as we have to allow for adding in the carry digits at each stage, but it is easily done with suitably connected logic gates. The addition circuit is usually presented as an MSI device. Other mathematical operations can also be performed by logic. For example, we can multiply two numbers together by adding the same number to itself repeatedly and counting the number of times we add it.

Programmable logic

Logic circuits can be built from a standard range of logic ICs of the kinds described above. All we need to do is to work out what gates are required and how to connect them together. Anyone who has attempted to design even a fairly simple logic circuit knows that the number of ICs required can easily mount to the twenties, thirties or even forties. These need a large circuit board, consume a large amount of power and require a complicated pattern of interconnections on the board.

For a mass-produced circuit of some complexity, such as a computer, it is possible to have specially designed LSI or VLSI circuits incorporating all the logic on one chip. But setting up production of a special IC is expensive and unlikely to be economic except for very large runs. Also, any change or improvement in the design requires the masks to be scrapped and re-designed. With programmable logic we use ICs that are produced in a few standard types, and so are inexpensive, but can be programmed to perform any of a wide range of logical operations.

One of the simplest forms of programmable logic IC is programmable read-only memory (PROM). These are normally used to store computer programs in a permanent form: for example, the system program that a computer uses when it is first switched on, to get it running. As an example, consider the simple PROM shown below, which has fewer inputs and outputs than usual, to make the diagram simpler.

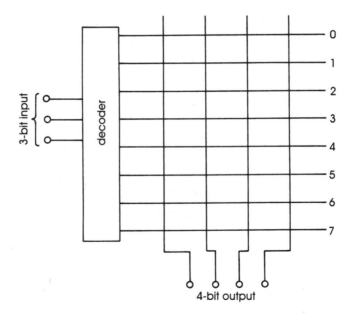

A programmable read-only memory works by linking some of the decoded input lines to some of the output lines.

This has three inputs and four outputs. The inputs go to a three-line to eight-line decoder. For each of the eight possible combinations of states of the inputs (000 to 111), one of the eight lines goes high.

There is a link between each of these lines and the four outputs, the link consisting of a transistor and a fuse. When the device is programmed, other circuits not shown in the figure allow us to decide which fuses are to be intentionally blown and which are left intact. This is what we mean by saying that the device is programmable. Suppose the input is 110, so that line 6 is made high. If all the fuses are present, base current is supplied to all transistors on that line and all four output lines go high; the output is 1111. But some of the fuses may have been blown when the device was programmed. If any fuse is blown the corresponding output stays low.

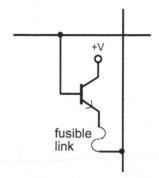

There is a transistor and fusible link at every crossing-point in the figure opposite.

Depending on which fuses are blown, the output corresponding to an input 110 can be any one of the sixteen 4-bit binary numbers from 0000 (all blown) to 1111 (none blown). Once a fuse has been blown it can not be repaired, so the device is programmed once and for all. After that, it is possible only to read the data stored in it, but not to change it, which is why it is called a **read-only memory**.

Used as a memory, the device is programmed with data which might be a table of information or might be the steps of a program in code. According to the 3-bit input (or address), we obtain an output corresponding to the stored data. The use of the PROM as a logic array is similar, as shown by the following example. An automatic lamp-switching system for a shop has three inputs: a passive infra-red detector in the store-room, a light sensor to detect if it is day or night, and a weekly timer to register if the day is Sunday or not. The output goes to four sets of lamps: the shop, an illuminated display in the shop, the store room, and the car park. The lamps are to be switched on as follows:

In the shop: every night, every day of the week.

The display: during day-time, every day, except Sunday.

In the store room: all the time, except day-time on Sunday.

In the car park: if a car is detected at night, except Sunday night.

The truth table for these requirements is:

INPUT			OUTPUT			
I-R (1=car)	Day/Night (1=night)	Day (1 = Sunday)	Shop lamps (1=on)	Display lamps (1=on)	Store lamps (1=on)	Car park (1=on)
0	0	0	0	1	1	0
0	0	1	0	0	0	0
0	1	0	1	0	1	0
0	1	1	1	0	1	0
1	0	0	0	1	1	0
1	0	1	0	0	0	0
1	1	0	1	0	1	1
1	1	1	1	0	1	0

The table has eight rows, covering all possible combinations of inputs. The outputs could be obtained by using a number of NAND and NOR gates, but blowing the fuses of a PROM to correspond with the 1's in the output columns of the truth table achieves the same effect with just a single IC and there is no need for connections between gates.

Sequential logic

The logic we have considered so far is **combinatorial logic**. We set up a number of fixed inputs to the circuit and the output takes a given value, depending upon the truth table or tables of the circuit. Given a certain combination of inputs, the circuit always behaves in the same way. In **sequential logic**, the behaviour of the circuit is partly or wholly dependent on what has been happening to it previously. A sequential circuit built with transistors is the flip-flop (or bistable) circuit shown on page 170. We give it a low input by connecting input A briefly to ground. What happens as a result of this input depends upon what state the circuit is in.

If output A is high, nothing happens. If output A is low, it changes to high. The behaviour of the flip-flop goes through a sequence of events. This is just a simple example of a sequential circuit but, from a simple flip-flop, we can build up more complicated circuits such as counters. The four outputs from a counter made from four connected flip-flops take one step through the sequence 0000, 0001, 0010, 0011, ... , 1111 each time the input to the first flip-flop is grounded. It counts from 0000 to 1111 in binary, that is from 0 to 15 in decimal. Sequential logic is important for the control of equipment such as an automatic drilling machine or a robot which has to perform a sequence of different tasks.

Logic circuits and the outside world

Logic circuits operate on electronically-represented facts but a logic circuit on its own cannot operate on any facts unless it has been told what they are. At the other end of the process, the circuit needs to be able to communicate the results of its operations to the user.

One of the most obvious and simplest ways of giving facts to a logic circuit is to open or close a switch. It might be a micro-switch attached to a door. When the door is open the switch is open; when the door is shut the switch is closed. The switch tells the logic circuit about the state of the door. But even such a simple arrangement has its problems. One of the major ones is **contact bounce**. It happens when we operate any switch, such as a key on a keyboard. As the key is pressed and the contacts come together, they do not necessarily stay together. They may make and break several times as the switch closes, closing together finally when the switch has reached the end of its travel. Contact bouncing may take as long as 20 ms.

This 'off-on-off-on-off-on-off-on' nature of a switch is unimportant when, for example, we are switching on a room lamp. The response of the lamp is slow, since it takes an appreciable time to heat up the filament to the temperature at which it begins to emit light. By the time it is heated the contact bouncing is over. The lamp does not appear to flash on and off. But logic circuits are fast, operating in microseconds or even nanoseconds. They have no difficulty in responding to every off-on change. If a logic circuit is being used to count how many times we press a key it counts once for every off-on change. We press the key once, but the counter may register 4 or more off-on changes.

The solution to this problem is to de-bounce the switch or key. The figure overleaf shows a debouncing circuit using a flip-flop (p. 170). Here the flip-flop is a logic IC, not made from discrete components, but its action is the same.

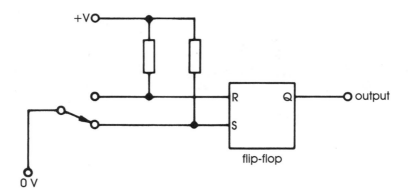

A set-reset flip-flop is used for debouncing a switch.

The two inputs, corresponding to inputs A and B on p. 170 are known as the 'set'(S) and 'reset' (R) inputs. When not connected to the 0 V line by the switch they are held high by the resistors. Connecting S to 0 V sets the flip-flop. Its output Q goes high. Connecting S to 0 V *again* has no effect, for the flip-flop is already set. Its output can be changed only by connecting R to 0 V, which resets the flip-flop, making Q go low.

In the figure above, the switch is connecting S to 0 V so the output is high. As we move the switch to its other position, contact bounce will cause makes and breaks between S and the 0 V line, but this has no effect on the flip-flop. However, the first time that the switch connects R to the 0 V line, the flip-flop changes state and its output goes high. Having done so, it is completely unaffected by any subsequent makes and breaks between R and 0 V.

One of the most important differences between a logic circuit and the outside world is the way in which quantities are represented. In the real world, temperature is a quantity which changes smoothly as an object is warmed or cooled. On a mercury thermometer we can see the end of the column moving smoothly up or down the tube. The thermometer scale may be marked in degrees but we can estimate fractions of a degree quite easily. The same applies when we use a thermistor. Temperature changes smoothly and so does the resistance of the thermistor. If the thermistor circuit represents changes in resistance as changes in voltage at its output, we find that the voltage changes smoothly too.

The way the voltage changes mirrors the way the temperature is changing. The rate and extent of the change in voltage corresponds to the rate and change of the temperature. We say that the voltage is an **analogue** of the temperature. In the same way, a microphone is sensitive to pressure changes in the air caused by sound waves. It generates an alternating voltage which we can amplify. The waveform of the amplifier output is an analogue of the waveform of the original sound.

By contrast there are no smoothly varying quantities in a logic circuit. Voltages change abruptly from 0 to 1 and 1 to 0, and in-between states are not recognized. A value is represented as a binary number which exists in the circuit as a series of highs and lows, representing the digits of the binary number. The value is represented in **digital** form. Informing a logical circuit about quantities in the outside world first involves converting it to an electrical analogue, usually a voltage, and then converting the voltage into digital form. Converting into an analogue is done by using a sensor such as a thermistor or microphone with a suitable circuit such as an amplifier to produce an analogue voltage. Converting this voltage into digital form is done by using an **analogue-to-digital** converter, or ADC.

analogue input   digital output

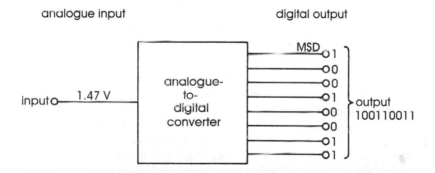

An ADC converts a real-world analogue signal into a digital signal that can be handled by a logic circuit.

The details of how an ADC works are too complex to go into here. An ADC is usually an integrated circuit which has a single input to which the analogue voltage is fed. It has a number of logic outputs, often eight or twelve, which may be high or low. The outputs produce an 8-bit or 12-bit binary number, its value depending on the voltage fed to the IC. If there are 8 bits, the output ranges from 0000 0000 to 1111 1111. In decimal this is from 0 to 255. There are 255 steps from the bottom to the top of the range.

For convenience (and also to simplify this explanation) we may set the analogue input to range over the values 0 V to 2.55 V. 2.55 V divided into 255 steps gives 0.01 V per step. If the input increases by 0.01 V, the output increases by 1. If the input is 1.47 V, the output is 147 in decimal, represented by 10010011 in binary.

With an 8-bit ADC as described above, while the input may vary smoothly, taking *any* value between 0 and 2.55 V, the output is allowed to take only 256 different values in steps of 0.01 V. The **resolution** of such an ADC is 0.1 V. This may be sufficiently good for many applications. A 12-bit ADC gives much better resolution since there are 4096 steps in increasing from 0000 0000 0000 to 1111 1111 1111.

After a logic circuit has performed its logic and its calculations it needs to make contact with the outside world again. Logic circuits can drive indicators such as LEDs directly, as well as seven-segment LED displays and LCD displays. Through the action of a transistor, a relay or an optocoupler, they can switch lamps, motors, solenoids and various other devices on or off. Sometimes a simple on-off action is not enough. The logic circuit may be required to control the brightness of a lamp or the speed of a motor, for example. In such cases the circuit must be able to produce a variable output voltage. This is usually done by using a **digital to analogue** converter, or DAC. The action of the DAC is the reverse of that of the ADC. It accepts an 8-bit or 12-bit binary input from the logic circuit and produces a voltage output of equivalent value. The output voltage is then used to control the brightness of the lamp, the speed of the motor or any other variable-rate process.

An alternative technique is **pulse width modulation**, which can be used to control the brightness of a lamp or the speed of a motor. The output of the digital circuit is a train of pulses of varying width. The actual voltage delivered to the lamp or motor is either high or low, that is, it is a binary signal. However, if the lamp is to be bright, the pulses are long ones with no gap or only a small gap between them. A motor will run at maximum speed. If the digital circuit reduces the length of the pulses and increases the gap between them, less energy is delivered to the lamp or motor. The lamp dims and the motor runs more slowly. In all cases, the pulses are of relatively high frequency, say 1 kHz or more, so that the lamp does not flicker and the motor does not run jerkily. In practice, this technique is better than an analogue voltage for controlling motor speed. The power when on is *fully* on, and this makes the motor less subject to stalling at low speeds.

20 Audio electronics

One of the principal aims of audio electronics is the true recording and reproduction of sound. Other aspects of audio electronics include the generation of musical sounds, a topic that is dealt with later in the chapter. In describing recording and reproduction we refer only to musical instruments, but most of what is said applies also to speech and song as well as other sounds and noises.

Tape recording

The magnetic tape recorder became available in the late 1940s and soon became very popular. The facility of being able to record at home without requiring additional or expensive equipment was a big factor in its widespread use. The principle of magnetic tape recording is illustrated overleaf. The polyester tape is coated with a layer containing a magnetic substance such as ferric oxide, chromium dioxide or pure iron in a finely divided form. The coating can be thought of as partitioned into a large number of minute regions known as **domains**. In any one domain, all the molecules of the material are arranged in one direction, so the domain is, in effect, a small magnet.

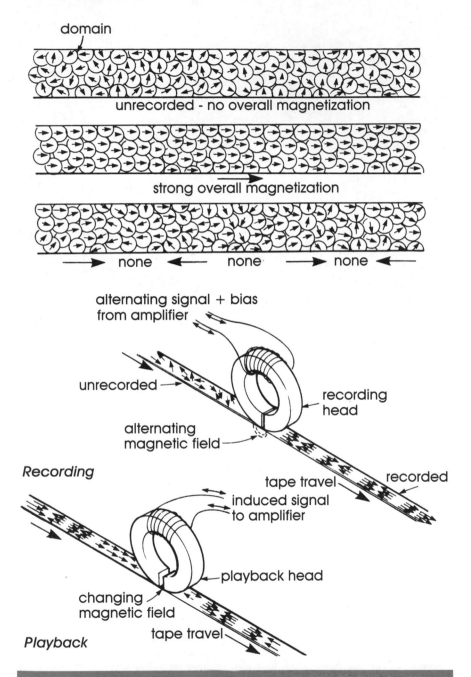

domain

unrecorded - no overall magnetization

strong overall magnetization

→ none ← none → none ←

alternating signal + bias
from amplifier

unrecorded →

recording
head

alternating
magnetic field

Recording

tape travel

recorded

induced signal
to amplifier

playback head

changing
magnetic field

tape travel

Playback

Magnetic tape recording depends on orienting the magnetic domains.

In unmagnetized tape the domains are magnetized in randomly different directions so there is no magnetic field around the tape as a whole. If the tape is fully magnetized (saturated), all domains are magnetized in the same direction. In a tape that carries a recording, an intermediate state exists in which a proportion of domains are aligned but the remainder point in random directions.

During recording, the tape is moved at a fixed speed past the recording head. This consists of an electromagnet which has a narrow gap between its poles. The signal from the amplifier is fed to the coil, so the magnetic field is an analogue of the original sound. The proportion of domains similarly aligned and the direction in which they are aligned depends on the strength and direction of the magnetic field as the tape passes the recording head.

On playback, the tape passes over a playback head. This is similar to the recording head and on all but the most expensive machines the same head is used both for recording and playback. But the action of playback is the opposite of recording. As the tape passes the head the field produced by its magnetization induces a varying current in the coil of the playback head. The varying current is an analogue of the original sound. Recording of stereophonic sound is easy with a tape recorder. It simply needs two heads, one for each channel. The heads are mounted one above the other as a single unit and produce two parallel tracks on the tape.

When a region of tape is magnetized during recording, the speed of the tape must be such that this region has moved away from the head before any significant reversal of the field occurs. The higher the tape speed, the better the quality of the sound. The standard speed for a cassette tape is 47.625 mm per second. This is half the standard speed previously used on open-reel recorders and is made possible by the improvements in tape and head quality made since the early days of tape recording.

Induction and motion

Induction occurs only when there is a *change* of magnetic field. The current induced in the playback head is the result of changes in the magnetic field as the tape passes the gap between the poles. If the tape is stopped, there is no change, no induction and no signal.

For playback it is essential that the head should have a narrow gap (about 1.2 μm) so that it is subjected to only a restricted region of the tape as it passes over the head. If a separate recording head is used, it may have a wider gap, perhaps 10 μm. The fact that the magnetic field may vary in strength and may reverse in direction while a point on the tape crosses the gap does not have any effect on the quality of the recording. This is because the magnetization of the tape depends on the field as the tape leaves the gap. Effective magnetization is confined to a narrow region of tape adjacent to the pole last passed by the tape.

One of the features of magnetic materials is that they do not become magnetized at all unless the magnetic field exceeds a certain minimum strength. This means that signals at low level are not recorded.

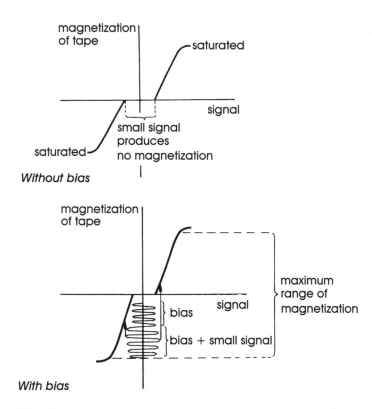

When there is no bias, a small signal fails to magnetize the tape. When bias is added to the signal, even the smallest signal will cause magnetization.

The audio signal is one that alternates and passes through zero twice during each cycle. The upper graph opposite shows what happens if an unmodified audio signal is applied to the recording head. We are looking at a part of one cycle of the signal as it swings from its negative peak through zero to its positive peak. At both peaks the magnetization is strong (more about that in the box) but when the signal is close to zero the tape is not magnetized. On playback, the signal has a step discontinuity as shown in the figure, which is a considerable distortion of its previous smooth form. The problem is overcome by the use of a.c. bias (lower figure). Before recording it, the audio signal is mixed by a high-frequency a.c. signal. This in effect brings the average strength of the magnetic field up into the range at which it is recordable. Since the a.c. signal is in the ultrasonic range (usually 90 kHz), it is not heard on playback.

Linearity

Magnetization is not proportional to field strength. Their relationship does not give a straight-line graph — it is not linear. At low levels of field strength, there is little or no magnetization (which is why bias is needed). At high levels, the tape approaches saturation. In between low and high there is a region in which magnetization and field strength are very close to being proportional but even in this region there is always a certain amount of distortion of the original waveform.

Erasing

One of the advantageous features of magnetic tape is that recordings can be erased and the tape used again for other recordings. It also allows corrections to be made to existing recordings. Erasing consists in disarranging the domains, so that they come to lie in random directions, as on an unrecorded tape. The technique is to subject the tape to a strong alternating magnetic field and gradually reduce the strength of the field to zero. Domains vary in the strength of field needed to alter their alignment. When the field is strong, all domains are affected at each change in direction of the field but, as the strength of the field is reduced, more and more domains become unaffected, each remaining in the direction it was in when it first became unaffected. Eventually all are randomly orientated.

Most tape recorders have a special erasing head. This is fed by the same a.c. signal as is used for bias, though at greater strength. The gap of the erasing head is large so that its field extends over a considerable length of tape. As a region of tape passes away from the head it experiences a gradually decreasing alternating field and is demagnetized.

The final stages

Returning to the drawing on page 226, we see that the playback of the tape is followed by amplification of the induced signal from the playback head to give it enough power to drive a loudspeaker. The cone of the loudspeaker is made to vibrate and this causes the air around it to vibrate in longitudinal waves, creating sound. The aim has been to make this sound as indistinguishable as possible from the original sound. Just as much depends on the later stages as on the early ones. In particular, the design of the loudspeaker and its enclosure play a major part, but these are mechanical aspects of sound reproduction that we can not go into here.

Similar considerations apply when a recording on one tape is copied on to a new tape. The original recording taken at a live session may be as nearly perfect as is possible. During the production of the final version of the recording there may be several stages in which tracks from different masters are mixed on to one tape. Multiple copies of the final version must be made for selling to the public. There is a loss of quality at every copying stage and the marketed version can never be as good as the original.

Before leaving this topic, it should be mentioned that the fidelity of reproduction depends on many factors other than those already mentioned. These include the smoothness of the tape surface, the type of magnetic material used, and the construction and materials of the tape head. There are also the consequences of age and wear. Tape stretches with age and its surface may become abraded. The magnetic heads eventually become worn and the gap width increases. These factors lead to progressive deterioration in sound quality.

Digital audio tape

Digital audio tape, or DAT, overcomes many of the defects of the analogue recording technique. The stages in the chain are the same as on page 226 with an ADC (p. 223) before the recording and a DAC after playback.

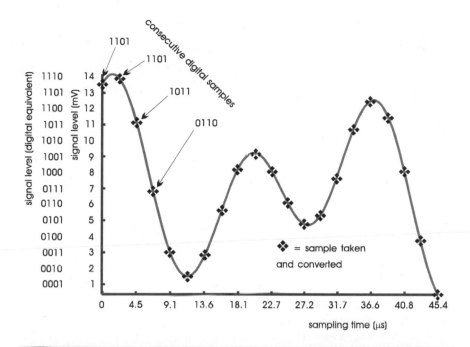

1101

1101

consecutive digital samples

1011

0110

❖ = sample taken
and converted

signal level (digital equivalent)

signal level (mV)

1110	14
1101	13
1100	12
1011	11
1010	10
1001	9
1000	8
0111	7
0110	6
0101	5
0100	4
0011	3
0010	2
0001	1

0   4.5   9.1   13.6   18.1   22.7   27.2   31.7   36.6   40.8   45.4

sampling time (µs)

For digital recording, a sample of the analogue signal is taken 41 000 times a second (usually) and each sample is converted to its digital equivalent.

From recording to playback we operate with a digital signal instead of an analogue signal and this makes a lot of difference to the quality of reproduction.

The graph shows the waveform of a very short segment of an audio signal. A sample-and-hold circuit registers the signal voltage 4100 times per second. The sampled analogue value is held long enough for it to be converted into a four-bit binary number by an ADC (4 bits are shown for simplicity – it is usually 12 or more bits). On the left of the figure the voltage scale is in millivolts and the digital equivalents. For the first four samples on the graph we have marked these digital equivalents. The digital equivalent of the analogue signal is 1101, 1101, 1011, 0110, ... , and so on.

The digital samples are a crude representation of the original. For example, the first two digital samples are equal though the voltages differ. There are two ways of improving it. One way is to take samples more often. Doubling the frequency of sampling would capture the first peak on the waveform.

The other way of improving the result is to have more binary digits (bits) in the conversion. In the example there are only four bits; this allows only 16 possible amplitude values. Given an eight-bit converter we obtain values ranging from 0000 0000 to 1111 1111 (0 to 255 decimal) making it possible to have 256 different amplitudes. With 16 bits, we can obtain 65 536, which is enough to reproduce the original waveform with extremely high precision.

It has been shown that to accurately reproduce the original waveform, sampling should take place at double the frequency of the highest frequency present in the sound. The highest frequency that is discernible by the human ear is about 20 kHz, so this is the highest frequency that we wish to reproduce faithfully. Accordingly, the analogue signal should be sampled 40 thousand times per second. This is a high rate of sampling but well within the capabilities of modern logic circuits. In practice, one of the standard rates for high-fidelity sound is 41 kHz, as shown on the diagram. The number of bits required depends on the application. An 8-bit ADC gives reasonable results with speech signals but, for high-quality music recording and playback, 16-bit conversion is preferred.

One of the prime advantages of a digital system is that, once a signal has been converted, it becomes a series of 0s and 1s, with no intermediate levels. In terms of tape recording, a zero is represented by a region of domains all magnetized in one direction. A '1' is represented by domains magnetized in the opposite direction. The tape is saturated in either direction. There are absolutely no problems associated with linearity, and no a.c. bias is required. Tape heads may be differently constructed and it becomes possible to record more data on a given length of tape. On reproduction, tape hiss due to non-uniformity of the magnetic particles and unevenness of the coating has no effect on the saturation of the tape, so is not heard on playback. When a digital tape is copied, 0s and 1s are copied as 0s and 1s; they do not change from one to the other. So a copy is an exact replica of the original. The effects of moderate tape and head wear have no effect on reproduction of digital signals. Thus many of the defects inherent in analogue tape recording are completely eliminated in digital recording. On the other hand, the digital signal can never be an exact copy of the analogue original. Digital signals are necessarily accurate only to nearest step, whether there be 16 such steps or 65 536 steps. The differences in each sample give rise to a type of noise (p. 189) known as **quantization** noise.

There is another way in which digital techniques bring advantages. We are dealing with digitally represented numbers which can be handled at high speeds by methods like those used in a computer. They can be **processed** in various ways.

When recording, there is no need to record the current sample on to the tape immediately. A whole batch of consecutive samples can be taken and stored in a memory. Various operations can be performed on them before they are recorded (as a batch) on tape. One of the processes is to take the original numbers and code them in another form to make them more suitable for recording. We can add so-called **check bits** to the numbers so that, if an error occurs at a later stage, it can instantly be detected. Sources of error include blemishes and drop-outs in the tape coating which can cause a temporary break or distortion of the signal at playback. During playback, the numbers recovered from the tape are decoded and analysed in batches, then sent to the DAC. If errors are detected, action can be taken to correct them, or at least minimize their effects. For example, if one sample is completely missing or (perhaps because of an error in one of its more significant digits) it is obviously too high or two low, the playback logic circuit can cancel that sample and substitute the averaged value of the two samples on either side of it.

Information recorded on the tape need not be limited to the sampled sound. Other data can be added as the recording proceeds. These include various codes to identify the batch number, the track number, number of tracks recorded, sampling frequency, and the time elapsed since the beginning of the recording. Information such as this enables the player to search for and play a selected track or tracks, or pick out a segment beginning and ending at a given time.

Compact discs

Compact discs, or **CDs** as they are more usually called, are digital recordings. The stages up to recording in the recording-reproduction chain are similar to those for digital, except that different systems of coding are used and the additional information added to the data stream is appropriate to a disc system. Samples from the left and right stereo channels are recorded alternately. This is possible because of the digital nature of the process. Among the many processes of coding and error-checking the digital data before it is finally recorded, we can shuffle batches of data so that the two channels, although both are converted to digital simultaneously, can then be stored, sorted and put on to the disc in alternate batches. The operation is reversed at playback and both channels are heard simultaneously.

The final digital signal is recorded on a disc of optically flat glass coated with a photo-resistant substance. Recording employs a beam from a low-power laser focused on to the underside of the disc (see overleaf). Although the laser has a power rating of only a few milliwatts its beam is focused on to a spot only 0.1 μm in diameter.

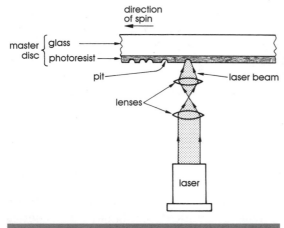

The intense energy of the beam at that point vaporizes the photoresist, leaving a pit about 0.6 μm in diameter and 0.1 μm deep. As the disc turns, the head produces a pit for every '1' in the digital signal, but leaves the disc unpitted when the digit is a '0'. The head is near the centre of the disc at the start of the recording and is moved toward the periphery as recording proceeds. Thus the pits are arranged along a spiral track with 41 250 turns.

During recording, the concentrated laser beam produces a pit in the photoresist for every '1' in the digital signal, leaving the disc unpitted for every '0'.

The rate of rotation of the disc is adjusted as the head moves outward, being gradually reduced to obtain a constant track speed of 1.2 m per second. Recording has to take place under scrupulously clean conditions as a single speck of dust could degrade several adjacent tracks.

A series of stages follow in which several copies of the master disc are made for use in mass-production. The discs eventually distributed in the shops consist of a transparent disc of polycarbonate, with dimples corresponding to the pits in the master. This disc is given a brightly reflective coating (usually of aluminium) which is then protected by having a further polycarbonate layer sealed over it.

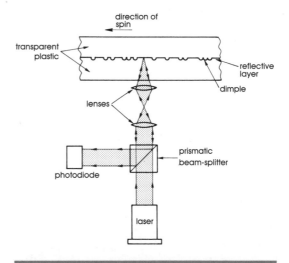

Playback works by detecting the reflection from the unpitted areas.

Playback uses another laser, of much lower power than that used in recording. The optical system projects a finely focused laser beam on to the disc. The beam is partly refracted by the transparent plastic coating of the disc. Because of this, the beam may be relatively broad where it enters the plastic layer; the plastic layer helping to concentrate it into a small spot on the reflective surface. Since the beam is broad where it enters the disc, the effects of small scratches on the surface are minimized. The beam strikes the reflective layer and, if no dimple is present, it is reflected back into the optical system. On its way back it is internally reflected in the prism, which directs the beam to a photodiode sensor. If a dimple is present, most of the beam is scattered sideways and little of it is reflected back to the diode. In this way the output from the diode consists of a series of low and high level representing the digital information stored on the disc. The signals from the photodiode are then sent to decoding circuits which check it for errors, strip off the various information (such as track number and elapsed time) and decode it into a straight digital analogue of the original sound. This is then sent to the DAC which produces an analogue audio signal ready for amplifying and feeding to the loudspeaker.

Various methods are used to keep the playback head centred on the track. In one of these, the three-beam system, there is a tracking beam to left and right of the main beam, each with its photodiode. One beam is slightly ahead of the main beam and one is slightly behind. If the main beam is correctly centred, the two side beams skim the sides of the dimples equally and receive signals of equal intensity. But if the main beam wanders off track, one or other of the side beams receives a stronger signal. The timing of the signals from the two beams tell the logic whether the main beam is too far to the left or too far to the right. Whatever the error, the positioning mechanisms of the playback head is adjusted to correct it. Another system is used to ensure that the focusing of the beam is continually adjusted to give a spot of the minimum size.

CDs share with digital tape the advantage that the recorded digital information remains unchanged no matter how many times the disc is played. There is no wearing of the grooves to degrade the reproduction as there is in a vinyl disc. But CDs are not indestructible and particular care must be taken not to scratch the outer surface of the plastic coating. The use of CDs as computer memory is described on page 247.

MP3

MP3 is a technique for compressing digital audio signals. This is more a matter of pure computing technology than it is electronic, but the popularity of MP3 stems from the development of flash memory (p. 245), which is electronic.

MP3 is short for *MPEG-1 audio layer-3,* and MPEG is short for the Moving Picture Experts Group. This international group developed a compression system for video images for use on the Internet and for digital TV. They have also produced a special technique for compressing audio files. If an analogue signal is sampled 44 100 times in a second and each sample contain 16 bits, this means that one *second* of playing time requires storage for 705 600 bits, or 88.2 kilobytes. Double this storage is needed for a stereo recording. Multiplying by 180 for a typical 3-minute popular track means 32 megabytes of storage. A Beethoven symphony requires considerably more.

The MP3 system uses the technique of **perceptual encoding.** Reasearch has shown that, when two sounds are playing simultaneously, we do not perceive the softer one. MP3 encodes only those sounds that we perceive and in conjunction with other compression techniques makes it possible to reduce the sound data to only 10% of its original length. This makes it possible to store up to an hour of music in the largest flash memory currently available. Just as important, it makes it possible for enthusiasts to download music from MP3 sites on the Internet without taking hours to complete the transfer.

MP3 tracks can be handled with an ordinary personal computer, given the appropriate software. Tracks can be downloaded from the Internet, or copied from pre-recorded CDs. MP3 has become so popular that many models of MP3 are available for the same purpose. Most of these have sufficient built-in memory for recording an hour or so of music, and most have plug-in flash memory cards that can be used for building up a collection of recorded music. MP3 players are small, portable, and battery-powered, playing into a headset or into a full-sized audio system.

The popularity of MP3 has raised much discussion regarding copyright, and has had a large impact on the music recording industry. As so often happens with new technology, development continues and now MP3 is beginning to be replaced with **Advanced Audio Coding** (AAC), which is 30% more efficient than MP3 at compressing music.

Electronic music

Electronic musical instruments fall into two main categories: **keyboards** and **synthesizers**. A computer with a sound card and suitable software can function in the same way as a keyboard. When a musical instrument plays a note it emits a sine wave of given frequency, the **fundamental**. At the same time it produces a whole series of **harmonics**, which are sine waves of higher frequencies: multiples of the fundamental.

The amplitude of the fundamental is usually greater than that of the harmonics and the higher harmonics usually have low amplitudes. The sound of a musical instrument is characterized by the harmonics that are present and their relative amplitude. The other feature that distinguishes the sound of a particular instrument is its **envelope.** Many keyboards are based on a number of oscillators running at the frequencies of notes of the musical scale. A selector switch is set to determine which instrument is to be simulated. Each time a key is pressed, logic circuits select the output from appropriate oscillators (for the fundamental and harmonics) and mix them (often using op amps) in their required strengths to make a sound resembling that of the instrument. Other circuits, using transconductance amplifiers impress the distinguishing envelope on the sound. The keyboard is also able to simulate percussion instruments, often by beginning with purely random waveforms (white noise) and filtering them and controlling their envelope to produce the required effects. An oscillator on the point of resonance may be used to simulate drums, gongs and guitar-like instruments. This is the method of sound production which is realistic for simulating certain instruments but not for others. There is also scope for producing pleasing *electronic* sounds which do not attempt to imitate any other musical instrument. Note that this technique of sound production uses analogue circuits.

Frequency modulation synthesis is a simpler and more often used way of producing sounds. In this, the fundamental sine wave is modulated by a second sine wave of higher frequency. That is to say, its amplitude is made to vary according to the modulating wave. This produces a note rich in harmonics. By adjusting the frequency of the modulating wave and also the depth of modulation it is possible to create sounds resembling recognizable musical instruments. However, the sound is more often interesting but is essentially 'electronic'. A more realistic way of simulating actual instruments is known as **sampled-wave synthesis** or **wave-table synthesis**. This is a digital method. The keyboard has a bank of memory in which are stored, in digital form, the recorded sounds of actual instruments being played. Only one note is stored for each instrument but when a note of given pitch is to be played it is processed by logical circuits to give it the correct frequency and to modify other characteristics of the sound.

Not only does the keyboard produce sounds but it has a battery of logic circuits to assist the performer to enhance the effects. With auto-chording, the player presses the keys singly but the instrument generates chords. The keyboard can also produce a rhythm accompaniment in a selected tempo, to back up the performance of the player. Keyboards have memory for storing performances. With these and other features the keyboard is a popular instrument, especially with the amateur player.

A synthesizer is an instrument for the serious musician. It comprises banks of oscillators, mixers, delay circuits, reverberation circuits and others with which the player is able to create an immensely varied range of sounds.

21 Computers

From the supermarket checkout to the flight deck of a jumbo jet, computers are playing an increasing part in our lives and all this has been made possible only by developments in electronics. Computer technology is one of the biggest contributions of electronics to changing, and we hope improving, the world we live in.

There are several kinds of computer, although the distinctions between them may be a trifle blurred. At one extreme are the **mainframe computers** operated by large companies and organizations. A mainframe computer usually occupies a room of its own and has a staff of operators. By contrast, a **microcomputer,** also known as a **personal computer** or as a **desktop computer**, takes up very little room and is operated by a single person. In general, both types of computer can do the same things; it is just that the mainframe computer has the ability to handle larger amounts of data. But the emphasis of development has been on microcomputers and most of them now have greater computing power than the earlier mainframes. Instead of having a company mainframe computer there has been a trend towards providing staff members with individual microcomputers linked to each other on a local network. In the private sector there has been an immense increase in microcomputer ownership, and this tendency is set to continue. **Lap-top computers** (see photograph) exploit the advances in electronic miniaturization and the development of large-area liquid crystal displays to provide a truly portable microcomputer.

The term 'computer' is not restricted to a machine with a keyboard and video monitor. Many other pieces of equipment incorporate computers to control their operations. These computers are dedicated to perform one particular task, such as controlling the operation of a clothes-washing machine. There are countless examples of such dedicated computers, found in satellites, guided missiles, CD players, Smartcards, and video games machines. There is more about these computers in the next chapter.

The main parts of a microcomputer appear in the block diagram opposite. The functions of the various parts are as follows:

System clock

This is usually a crystal oscillator (p. 175) running at the rate of several tens or hundreds of megahertz (tens or hundreds of millions of cycles per second) to set the pace for the activities of all parts of the system. The fastest PCs run at 1 gigahertz (a thousand million cycles per second).

Control unit

Taking its cue from the clock, and its instructions from the central processing unit, the control unit is a set of logic circuits which coordinates the activities of the system.

Central processing unit

This is where calculations and logical operations on the data supplied to the computer, are performed by binary logic. The CPU consists essentially of a VLSI integrated circuit known as a **microprocessor**. This contains several sets of registers, each comprising a set of 8, 16 or 32 circuits of the flip-flop type. These can be loaded with data in binary form. The data may consist of numbers, coded information, instructions to the microprocessor received from memory, or addresses of locations in memory. In some types of microprocessor, one of these registers, the accumulator, holds the data on which operations are currently being performed. In the accumulator the data may have other data added to or subtracted from it. The value held may be tested by comparing it with a given fixed value and subsequent action taken according to the results of the test. The data bits may be shifted along in the register in various ways. Individual bits may be tested to find out if they are '0' or '1' and appropriate action taken. These operations are all very simple in nature. The microprocessor does not in fact do anything very complicated. Its great power lies in its ability to perform many simple steps one after another at exceedingly high speed (millions of steps each second) without making mistakes.

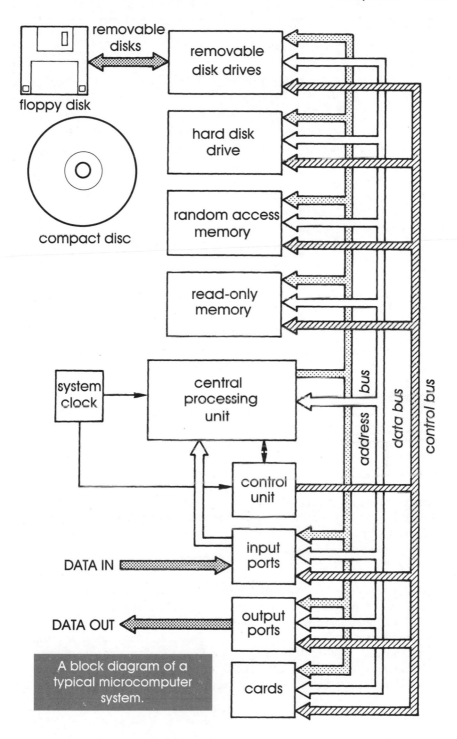

removable disks

floppy disk

compact disc

removable disk drives

hard disk drive

random access memory

read-only memory

system clock

central processing unit

address bus

data bus

control bus

control unit

input ports

DATA IN

output ports

DATA OUT

cards

A block diagram of a typical microcomputer system.

Addressing

The memory of a computer consists of millions of sets of flip-flops or similar circuits. We need a way of locating one of these sets individually, either to read data from it or to store data in it. Like the houses in a street, each set is identified by its **address**. An address consists of 16 (usually, but possibly more) bits. With 16 bits, we can address up to 2^{16} ($= 65\,536$) different sets of flip-flops, but the memory of most computers is much larger than this. The addressing range can be increased by increasing the number of address lines or by working with separate blocks of memory which are switched on or off as required. Addresses are also allocated to input and output ports for use when receiving or transmitting data.

Input ports

These receive data from peripheral devices such as a keyboard, a mouse, or a modem. In industrial and scientific applications, input may come directly from electronic sensors or measuring circuits. An example is given on page 231 showing how a varying voltage may be sampled and then converted to binary form by an analogue to digital converter. The digital data is read in by the computer and displayed on the screen to resemble an oscilloscope display. There is more about input peripherals on page 248.

Output ports

These transmit data to peripheral devices such as the video monitor and printer. Industrial computers may be connected to machinery (for example, a drilling machine) under their control.

Buses

These consist of sets of connecting lines (cables or tracks on a PCB) running in parallel. An address bus consists of 16 or more lines, each carrying one bit of a 16-bit (or longer) address. The advantage of having parallel lines is that all the bits of an address can be sent at once, which is much quicker than having to send the bits one after another along a single line. The address bus goes to all parts of the computer with which the CPU needs to communicate. When an address is put on to the bus by the CPU, it is received by all the devices connected to the bus.

The address is decoded in each device (for example, in a memory IC) and, if the address corresponds to a memory location in that device, that location is put into contact with the data bus. The data bus, which may have 8, 16 or 32 lines, is used for sending data between the various parts of the computer. Data buses are bidirectional; they may be used by the CPU to send data to memory to be stored or to read data that is already stored in memory. The control bus links the control unit with other parts of the computer.

The data bus is communal; each part of the computer can take its turn (under the direction of the control unit) to send or receive messages along it, but only one device can be allowed to use the bus at any one instant. The control unit must allow only one device to put data on the bus at any one time. In this respect, the control unit is like the leader of a discussion group allowing only one person to talk at one time. But the situation is more complicated than this. Even when a device is not active, its outputs are at logical '0's and '1's. The outputs of all the inactive devices, if connected to the bus, would interfere with the output of the one active device currently using the bus. We must be able to disconnect inactive devices. To do this, we use a **three-state output**.

The output of the gate is controlled by an 'enable' input, connected to one of the lines of the control bus. The input level (high or low) is passed through an INVERT gate in the IC, so that both the true (E) and inverted states ($\overline{E}$) are available. If E is high, $\overline{E}$ is low, and the output is enabled.

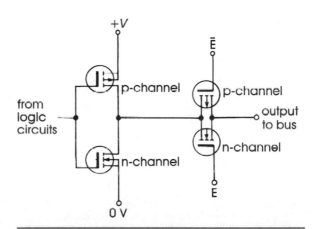

A CMOS three-state output can be completely disconnected from the bus.

This is because both transistors of the complementary pair are conducting. Then the output of the gate is high or low, depending on the logic before it. It behaves just like an ordinary gate, sending data to the bus. When the device is disabled, E is low (and $\overline{E}$ is high). Both transistors are off. This completely disconnects the output terminal from the rest of the gate, so it is unable to place data on the bus.

Interrupt lines

A signal, usually a low one, on one of these lines causes the microprocessor to complete whatever operation it is engaged in and then jump to a special program known as an **interrupt servicing routine**. The interrupt signal comes from an external device, such as a modem which has just received a message over the telephone system. The modem has received data that it has to send to the computer. In this case, the interrupt servicing routine causes the microprocessor to accept the data and perhaps store it in memory. Then, depending on the situation, the computer might resume what it was doing before the interrupt. Interrupts are handled so rapidly that the user would probably not notice that anything unusual had happened. Then, at some more convenient time, or when asked to by the user, the computer looks for the stored message and displays it on the screen.

Read-only memory

The computer has a small portion of its memory, called ROM for short, which usually contains the program data that is needed as soon as the computer is switched on, to initialize the computer and make it ready to receive instructions from elsewhere. In specialized computers, the ROM may contain longer programs, such as word-processor programs, which are therefore ready to run as soon as the computer is switched on. It is not possible for the computer to store data in ROM, but it can read data that has already been stored there, usually by the computer manufacturer. When the address of a given memory location is fed to the address inputs of a ROM chip, the outputs of the ROM go high or low, according to the data stored at that location. The CPU simply has to step through the addresses in order and a succession of instruction codes is put on the data bus to tell the CPU what to do next.

Data may be put into ROM by using special masks when making the memory chip or, with a different type of ROM chip, it may be programmed into it at a later stage, before it is installed in the computer. Data written into the ROM during manufacture is permanent and is available every time the computer is switched on. Data written into a programmable ROM (or PROM, p. 218) is also permanent in the sense that the fusible links can not be re-formed once they have been blown. There are also electrically programmable ROMs in which data is stored by placing an electric charge on the gates of MOSFETs. If a gate is uncharged, the output of the gate is low; if it is charged, the output is high. Each gate is electrically insulated so that charges leak away extremely slowly. Data stored in such a ROM remains for years, so it is permanent for all practical purposes, unless there is reason to change it. There is a quartz window above the chip to allow the chip to be exposed to ultra-violet radiation, which removes the charges and destroys the stored data in a few minutes.

This allows the ROM to be re-programmed, perhaps with an improved version of the program. In another form of PROM, known as **electrically eraseable PROM** (or EEPROM) the stored charge may be removed by electrical means. In EEPROM, the memory cells are configured in groups of 8, so that data is written and read a byte at a time.

In a variation of EEPROM, known as **flash memory**, the whole memory or large blocks of it can all be erased at a single operation. This is how this type of memory gets its name, because it is written and read 'in a flash'. Its function is similar to that of a hard disk drive (p. 247) because it can store data indefinitely, without requiring a continuous power supply yet its design lets the data be changed very rapidly when the need arises. Unlike a disk drive, it is a solid state device and has no moving parts. Flash memory is particularly suited to portable equipment such as digital cameras, mobile telephones, and MP3 players which need to store large amounts of data quickly. Flash memory may be built in to the equipment but it is also available as flash cards about the size of a large postage stamp and less than 2 mm thick. These have a memory size of up to 64 megabytes. An 8 Mb card can store up to 50 full-colour photographs of medium resolution.

Random access memory

The CPU can both read data from this type of memory (RAM) and write data into it. Essentially it consists of a number of flip-flops that can be set (output '1') or reset (output '0') by writing appropriate data into them. The flip-flops are arranged in groups of eight, each group having its own address, and providing storage for an eight-bit binary number or code. A typical microcomputer has several megabytes (million bytes) of RAM. The RAM is the workspace of the computer. It may be used to store the program that the computer is running, to store data on which it is to work, or to act as a 'scratch-pad' in which it can temporarily store the results of calculations. The essential point about RAM is that the CPU can store or read data at very high speed. This is essential if it is to be able to make full use of its ability to calculate and perform logic at high speed. However, there is a limit to the amount of RAM that a computer can conveniently use; 64 megabytes is the upper limit on many computers. This is partly due to the constraints of addressing a large memory and also the fact that having many RAM chips takes up space and requires extra power. This is why a computer needs additional storage, in the form of magnetic discs.

RAM differs from ROM in that the data stored in it is not permanent. All data in RAM is lost when the power is switched off. Any data that is to be retained must be transferred to other storage devices before the computer is switched off.

Cache memory is a special type of RAM with very short access time. It might be a special chip or it may even be located on the microprocessor chip. For example, the computer shown in the title photograph has 64 kilobytes of cache memory in the microprocessor and an additional 512 kilobytes in RAM. The purpose of cache memory is to increase the speed of processing. It is used for storing data that *might* be required in the near future. In principle, it is like keeping frequently-used telephone numbers in a short list close by the telephone. When data is needed, the CPU first looks in cache memory for it. If it is not there, it looks in the main RAM. **Video RAM** is a special area of about 2 megabytes of RAM set aside for storing the picture currently being displayed by the monitor.

Disk drives

Most computers have three kinds of disk drive. The **floppy disk drive** stores data on a thin flexible plastic disk coated on one or both sides with a magnetic film. Although the disk itself is floppy, and early ones were enclosed in flimsy cardboard covers, most disks are nowadays enclosed in a stiff plastic cover. The cover has a metal shutter which slides back automatically when the disk is inserted in the drive to expose part of the disk surface to the magnetic head.

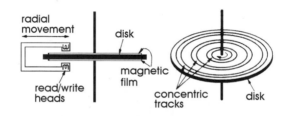

radial movement   disk

read/write heads   magnetic film   concentric tracks   disk

A magnetic disk drive (floppy or hard) has a dual read/write head, one for each side of the disk.

The principle is the same as that used when recording music on digital audio tape. The main difference is that the data is recorded on 40 concentric tracks and the magnetic head moves radially to read or write each track. Each track is divided into sectors, each one being allocated to one particular program or set of data. Longer programs or data tables may require more than one sector. There is a directory track on the disk telling the computer in which track and sector to look for each block of stored data and the magnetic head can skip from track to track and sector to sector, finding the information that is required. A typical floppy disk can store up to 1.4 megabytes of data.

Data can be read at rates of several hundred bits per second, but first the disk must be accelerated up to full speed (360 r.p.m.), and the magnetic head moved to the correct track and sector. A typical access time is 200 milliseconds, which is much slower than the 25 to 150 nanosecond access time of RAM or ROM.

A **hard disk drive** has one or more disks attached to the same spindle. The disks are made of non-magnetic metal and coated on both sides with a magnetic film. The principle of storage is the same but the magnetic heads are much closer to the film. This is because the disks rotate at very high speeds (about 3600 revolutions per minute). This gives rise to a thin layer of moving air close to the disk surface in which the magnetic head 'floats' without actually coming into contact with the disk. Since the head is closer to the disk, it is possible to record data more densely: the tracks are closer together and the recorded bits closer together than on a floppy disk. Consequently, a typical hard disk drive stores several gigabytes (thousand million bytes). Another advantage of the hard disk is that the high rate of spin reduces access time to about 20 milliseconds. As the head is very close to the surface of the disk it is essential to exclude particles of dust or smoke. Hard drives are sealed during manufacture and can not normally be opened by the user.

Compact disc drives are very similar to CD players and work on the same principles. In fact, they are able to play ordinary music CDs through the sound card of the computer. The information stored on a CD is simply a series of 0s and 1s. It may represent musical sounds but it could equally well be used for storing information of other kinds. In computing terms, a CD stores about 600 megabytes of data. CDs have largely replaced floppy disks as a medium for distributing software. Most programs nowadays are too long to fit on a floppy disk and there are other advantages too. A CD is unaffected by stray magnetic that can so easily wipe the data from a floppy disk. Also, CDs cost much less to produce in quantity than floppy disks, so are ideal for large-scale distribution, on the covers of computer and other magazines, for example.

Like hard disk drives, CD drives are fast enough to be used as memory storage devices for computers, the data being accessed straight off the CD. The main difference is that CDs are read-only memory (CD-ROM). However, **CD-recordable drives** may be used with special CD-R discs to write (but not re-write) data and play it back as many times as required. CDs are widely used in **multimedia technology.** A disc can store text, computer programs, photographs and diagrams, motion pictures and sound. These can be accessed and loaded into the computer almost instantly. Very elaborate games with startling graphics are now available on CD-ROM but more serious applications of the technology include educational and reference discs.

Cards

Most computers have sockets for plugging in optional circuit-boards (known as cards) to extend the capabilities of the system. These may include a **modem card** for communicating with the telephone network, a **fax card** to allow the computer to send and receive faxes, and a **sound card** to add sound producing and recording facilities to the computer. Another type of card is the **memory card**, which carries a number of RAM ICs, enabling several hundred kilobytes of data to be stored and read. Memory cards have a built-in lithium battery which powers the ICs whenever the card is removed from the computer, so that the data are not lost. **Flash cards** (p. 245) are replacing ordinary memory cards as they are smaller and cheaper, are able to store much more data, and do not need their own power supply.

Input peripherals

A computer needs one or more peripheral devices for communicating with the world around it. Some are concerned with input, others with output.

Keyboard

Most computers have this peripheral as standard, usually in the shape of a typewriter QWERTY keyboard, with a numeric keypad on one end and various other function keys. Each plastic key-top is held up against the plastic frame by a spring, with guides to keep it in position. There is a double plastic membrane beneath the frame, one with parallel tracks corresponding to the rows of keys, the other with slanting parallel tracks corresponding to the columns of keys. When a key is pressed, a track in one layer is brought into contact with a track in the other. The computer is programmed to scan the rows and columns repeatedly whenever it is waiting for input. If it detects a contact between a given row and a given column, it knows which key has been pressed. On page 221, we mentioned contact bounce, which is always a problem with fast-acting logic circuits. Debouncing each of the hundred or more keys of a computer keyboard by the method described there is not practicable, but a 'software solution' is given on page 254.

Mouse

Users who are not familiar with a keyboard may find a mouse a faster and more convenient way of telling a computer what to do. A mouse consists of a plastic box shaped so as to fit conveniently under the hand. The 'tail' of the mouse is the lead by which it is connected to the computer.

A plastic ball projects slightly from the under-side of the mouse and rests on the surface on which the mouse is placed — usually a plastic 'mouse pad'. When the mouse is moved, the ball rotates. Inside the mouse are three plastic wheels in contact with the ball, two of which rotate around perpendicular axes to detect left-right and up-down motion on the mouse pad. On the same spindle as each wheel is a sectored disc. Light from an LED passes between the sectors and is detected by phototransistors. The pulsed signal produced as the discs rotate indicates the rate of motion of the mouse in the two directions. This is interpreted by software and used to position the cursor on the monitor screen. The mouse has up to three buttons (only two are generally used) which close microswitches. These are used to signal instructions to the computer, for instance to indicate that the cursor is now on a part of the screen selected by the user. When the button is pressed ('clicked') the computer performs an appropriate action.

Trackball

This has the same function as a mouse but consists of a ball rotated by the fingers in a socket at the front of the keyboard. It does not require as much desk space as a mouse mat, so it is ideal for use with lap-top computers.

Digitizing pad

In one type of pad, a grid of wires is embedded in a rectangular plastic tablet. The computer sends a pulse to each horizontal wire and each vertical wire in turn, thus scanning the whole pad at high speed.

The pulses are detected by a sensor (such as a Hall-effect device) at the end of a stylus and a pulse is sent to the computer. By this means the computer can determine the exact location of the stylus tip on the pad. It can be programmed to display a dot at a corresponding location on the computer screen. As the stylus is moved, a line is drawn on the screen.

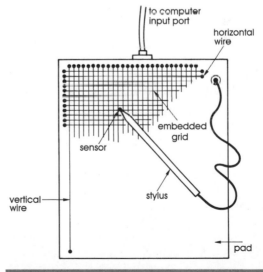

A digitizing pad for entering drawings.

Digitizing pads are used for entering graphical data. Rough drawings can be tidied by software and processed into a form suitable as artwork for publication.

Touch screen

The monitor screen has a shallow frame in front of it in which are located an array of ultrasonic or infra-red sources and sensors. When a person touches the screen, placing the finger on a displayed panel, signals from the sensors indicate the finger position. Depending on the position of the finger, the computer takes appropriate action. Another method uses a screen coated with transparent conductive strips. The capacitance between the finger and one of the strips indicates where the finger is pointing. In general, touch screens are limited in their usefulness by their low resolution. However, they are often found in displays in shops and museums. The display allows members of the public to select from a few displayed 'buttons'. This is easier and less daunting to persons who are unfamiliar with using conventional keyboards.

Scanner

A document such as a photograph or a typewritten sheet is scanned by moving a spot of light over it. Variations in the amount (and in a colour scanner, the colour) of the light reflected from the document are read by the computer and used to build up an image of the document on the screen. A colour photograph can be scanned and then printed out, or sent by e-mail. After scanning a text document, such as a page from a book, the image can be processed by software to convert it to a word-processor document with remarkable accuracy. Scanning a document into a computer and correcting the few inevitable errors is much quicker than copying it by typing on the keyboard.

Magnetic stripe reader

The resistance of ferromagnetic materials depends upon the strength of the local magnetic field. In a **magneto-resistive sensor**, a current from a constant-current generator passes along a strip of ferromagnetic material. As the local field alters, so does the resistance of the strip and therefore the p.d. across it. The variations of p.d. can be measured. Inductive magnetic sensors depend upon the *rate of change* of magnetic field, but magneto-resistive devices depend solely on *field strength*. These sensors are used for reading data recorded magnetically on tapes and computer disks. They are also used for reading from the magnetic stripes on bank cards, tickets, and identity cards. Because they do not depend on the rate of change of field strength, they are unaffected by the rate at which the magnetic stripes are swiped through the reading machine.

Optical disks

Although computer data is most often stored on magnetic disks, other forms of data storage may sometimes have advantages. One of these is the optical disk, in an optical disk drive. An optical disk consists of an active layer sandwiched between two transparent layers. Data is recorded on the disk by directing a high-powered laser beam on to its surface. The beam heats and melts a small spot on the disk to correspond with a '1'. With one type of active layer, the material crystallizes as it cools. This changes the reflectivity of the surface, producing a series of spots that are more reflective than the unmelted material. The disk is read by directing a low-power laser beam on to it and detecting the amount of light reflected. Such a disk is a 'write once read many' or WORM disk.

Another type of active material can be used which is already crystallized and has high reflectivity. A high-power beam is used for writing, destroying the crystalline structure and so producing spots with lower reflectivity. This can be read using a low-power laser beam, as above. However, if the writing beam has medium power, this is sufficient to restore the crystalline structure and can be used to erase data already written on the disk. As the writing laser scans the disk, it can be energized at high power where it is to produce a new spot or at medium power where it is to erase a previously recorded spot. Thus the disk can be overwritten with new data.

Magneto-optical disks

Magneto-optical disks in a suitable drive are another technique used for storing computer data. They store data as a pattern of bits by directing a laser pulse at the magneto-optical material in a magnetic field. The laser beam melts the material, the molecules of which orientate themselves in the field. The material immediately cools, leaving the small spot permanently magnetized, representing a logical '1'. The disk is read by using a polarized laser beam. Magnetization alters the reflective properties of the surface of the disk. As a result of this, the direction of polarization of the beam is changed if it is reflected from a magnetized spot. This change is detected and used to distinguish between magnetized (= 1) and non-magnetized (= 0) regions. Such a disk can be written to repeatedly to alter the stored data.

One limitation on the amount of data that can be stored is the size of the spot that can be heated, in other words the diameter of the laser beam. Normally lasers emit radiation in the red region of the spectrum, where wavelengths are longer, and the minimum beam diameter is 0.8 μm. Increased data density is obtained by using a blue laser which can be focused to a spot only 0.4 μm in diameter.

A magneto-optical disk the same diameter as a CD can store up to 6.5 gigabytes when written to by a blue laser. This is enough room for 6500 novels each 250 pages long. Data may also be stored on miniature magneto-optical disks that are only 2.5 inches in diameter and which hold up to 140 megabytes.

Output peripherals

The most important output peripheral is the monitor, or visual display unit. This may be of the TV type, using a variant of the cathode ray tube, or it may be a liquid crystal display. The principles on which these work have already been described so we will say no more here. Other output peripherals include loudspeakers of better quality than those normally built in to the computer itself. A device that has both output and input functions is the **removable hard disk drive**. The drive uses hard disks that are slightly larger than floppy disks, but store up to 250 megabytes of data. That is the equivalent of 174 floppy disks. Reading and writing is about the same as for a hard disk. This combination of the large capacity and fast transfer of the hard disk with the convenience of the removability of floppy disks make these disks ideal for backing up, archiving and large-scale data storage.

Printers

Thermal printers have a head that consists of a row of tiny heating elements. The paper has a special coating that turns dark grey when it is heated. By turning appropriate elements on or off as the paper passes the head, the computer can print characters and designs. Thermal printers are inexpensive and silent in operation. They are often small, printing out on paper tape a few centimetres wide. They are often used in supermarkets and shops for printing out the bill. Their main disadvantage is that the printing fades with time and may be obliterated by excessive heat.

The head of a **dot-matrix printer** consists of a vertical row of nine fine needles. A coil surrounds the base of each needle and the needle is propelled toward the paper when the coil is energized. As the head travels along a horizontal track, the needles impact an inked ribbon, causing the paper behind the ribbon to be marked. The head scans across the paper, printing a row of characters. Then the paper is moved on to print the next line. Dot matrix printers are usually limited to printing in black. They can print on almost any paper, and on separate sheets, though continuous perforated paper edged with sprocket holes is commonly used. Their main disadvantage is that they are very noisy. Under computer control, they can be made to print in a wide range of fonts and can also be programmed to print simple graphic designs.

At one time, dot-matrix printers were the commonest kind of printer used in offices, but more recently they have been replaced by the inkjet printers. The head of an **inkjet printer** works by squirting microscopic drops of ink at the paper as it scans across the sheet. These printers are very quiet in action, though not silent, are reasonably inexpensive, and have the big advantage that they are able to print in full colour. They print on separate sheets and on any paper surface, though special paper with a gloss surface is best for printing colour photographs.

The head of the printer is perforated with between 200 and 1000 pinhole nozzles, about 70 μm in diameter. They are in staggered rows spaced only 80 μm apart so as to give the printing a high resolution of 600 or more dots per inch. For colour printing, the nozzles are in groups supplied with inks of four different colours. The inks are coloured cyan (blue-green), magenta (red-blue), yellow and black. The first three are the subtractive colours that, when mixed in varying proportions, can produce all the colours of the spectrum in all shades from pale to dark. It is preferable to print true black using black ink, rather than using a mixture of all three subtractive colours.

Behind each nozzle is a compartment filled with ink (A). At the rear is a thin wire which acts as a heater. To eject a drop, a strong but short pulse of current is passed through the wire (B). The heat causes a bubble of vapour to appear around the heater. The extra volume of the bubble produces a pressure wave which forces a jet of ink out through the nozzle. After the pulse the bubble contracts (C) and a tiny drop of ink breaks away from the jet. The drop rounds itself off (D) until it is a perfect sphere as it hits the paper. All of this takes place at high speed, so that about 12 000 drops come from the nozzles of the heater in one second. In some printers, up to four drops can be directed at the same point on the paper. This is equivalent to having drops of variable size, giving a graduation of the tone of each pixel.

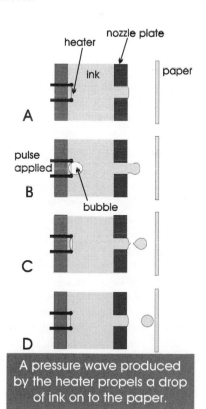

A pressure wave produced by the heater propels a drop of ink on to the paper.

Another way of producing the pressure wave uses a small piezo-electric crystal. Behind each nozzle, the wall of the ink compartment forms a thin diaphragm. The crystal is held against this. When an electrical pulse is passed through the crystal, it changes shape, pressing on the diaphragm and so producing a pressure wave in the ink and ejecting a drop from the nozzle.

Hardware and software

We have already mentioned software in passing but now we look at the concept of software more closely. When talking about a computer system, engineers divide it into two major parts: hardware and software:

Hardware refers to the physical units such as the central processing unit, the memory, or the drives that make up the computer.

Software, on the other hand, is the term used for the programs that control the operation of the computer through the CPU and enable calculations, logical operations, fault detection and other functions to be performed automatically.

Software is usually stored on magnetic disks, mostly on the hard disk, which has ample room for long programs. When a program is to be run, the software is copied from the disk into RAM. Some programs are so long that there is not room for the whole program to be held in RAM at any one time. In such cases, different parts of the program are transferred to RAM from the disk as and when they are required. Conversely, if RAM holds a lot of stored data that is not currently required, it can be transferred to the disk and re-loaded back into RAM later. All of this swapping between RAM and the disks is performed automatically according to the operating system of the computer.

Very often it is possible for software to take over the functions of hardware. An example of this is the debouncing of keys on the keyboard. This could be done by having a capacitor and inverting gate connected to each key, but software offers a much simpler solution. When the computer detects a key press, it registers the fact but does nothing more immediately. At intervals of a fraction of a second it repeatedly scans the keyboard, registering the key-press each time. After it has found the key to be pressed on several consecutive scans, it accepts the fact that the key is pressed and takes appropriate action. This routine can also be used for avoiding the effects of rollover, when a key is pressed before the previous one is released so both are held down at the same time. The computer can decide which key was pressed first and which next.

We referred above to the **operating system**. This is the software that instructs the microprocessor on how to perform the essential tasks of running the computer. The initial step, the **BIOS,** is stored in a special area of ROM.

BIOS is short for **Basic Input-Output System**, and this controls the flow of data within the system. Because it is provided by the manufacturer of the computer as a pre-programmed ROM, it is often referred to as **firmware**. During the initial stages, the microprocessor is told where in memory to find the software of the main operating system. This is usually stored on the hard disk drive, though it may be on a floppy disk, or in ROM. The microprocessor checks that memory is working properly and finds out how much memory is available. It checks the input and output ports to discover what devices are attached to the system. It does countless other tasks intended to make the computer ready to do what the operator rquires.

The main part of the operating system is stored on the hard disk and is concerned with transferring programs from the disks and running them. It also looks after the display on the monitor screen and the acceptance of commands from the keyboard. It includes a number of programs that the user can use to load and store data, save blocks of data (files) on disk, load blocks of data into RAM, send data to the printer, make copies of disks and many other useful tasks. Originally, this collection of software was called the **disk operating system**, and known as **DOS** for short. Later, manufacturers developed a number of different and improved systems for different computers. The most popular of these is **Microsoft® Windows®**, which over the years has appeared in a succession of expanded versions such as Windows® 95, Windows® 98, and Windows® 2000. However, there are a number of other operating systems such as Unix (used on minicomputers) and Linux which have certain advantages.

Languages

Microprocessors are instructed what to do by a sequence of coded instructions stored in ROM or RAM. The code may be an 8-bit one or longer. The microprocessor goes to the address where the first instruction is stored, and reads it (by way of the data bus) into its input register. From there, the logic of the microprocessor interprets the code and the microprocessor performs the required operation. Having read and executed the instruction, the microprocessor reads the code from the next byte in the sequences. So it goes on, working through the sequence and occasionally being instructed to jump to other parts of the sequence, or to jump back to an earlier part and repeat a section of the program. The program that is stored in RAM or ROM as a succession of bytes (or larger units) is known as **machine code**. Machine code is the only kind of instruction that a microprocessor 'understands'. If we want to communicate with the microprocessor we must use machine code.

Unfortunately, machine code is difficult for humans to understand. Writing a program in machine code is a tedious matter and it is all too easy to make mistakes. For this reason software has been written to make things easier. The simplest kind of software is known as **assembler**. Using an assembler, the programmer types in the instructions using a set of **mnemonics** instead of the digital codes. The mnemonics are intended to be easier to remember than machine code. For example, 'LDX 45' might mean 'load register X (in the microprocessor) with the value 45'. When the programmer has finished, the assembler automatically converts it into machine code.

High-level languages are even easier to use, at least, for programming simple operations. These allow the user to key in the instructions in a form that more closely resembles real language, usually English. The most popular of these is *BASIC*. The user types in instructions using intelligible statements, such as 'LET temperature = 45'. This assigns the value 45 to a variable referred to as 'temperature'. Even an operation as simple as this would take several steps if written in machine code. *BASIC* includes an **interpreter** program that reads the stored program, converts it to machine code for immediate use by the microprocessor. Because the microprocessor has to use the interpreter every time the program is run, *BASIC* tends to be slower than other high-level languages. However, with the high clock speeds of modern computers the program still runs at lightning speed.

Other high-level languages include *C*, *FORTRAN*, and *PASCAL*. One of the newer ones is *JAVA*, which has the advantage that it can run on many different kinds of computer. This makes it useful for writing the short programs that produce the scrolling messages and other effects that are common (and distracting!) on pages of the Internet. Each high-level language has its own special syntax and each is particularly suited for certain types of program. These languages have a **compiler** program that turns original code into machine code when the program is complete. From then on, the computer runs the compiled version, which is the fastest form of the code.

Special 'visual' versions of some of these high-level languages have been produced for programming under the Windows® operating system. *Visual BASIC*, *Visual C* and *Visual Java* are examples. They include a wide range of commands especially for formatting the Windows® displays and enabling them to interact with the user.

22 Microcontrollers

A microcontroller has been described as a 'computer on a chip'. This is an oversimplification, for a microcontroller lacks much of the hardware, such as the disk drives, monitor and keyboard of a computer but, at the same time, it is more than just a microprocessor.

A microcontroller is a single VLSI chip which holds not only a central processing unit (CPU) but a number of other units too. The CPU is, of course, a microprocessor. This is usually simpler than the microprocessors found in microcomputers but it is more than adequate for the tasks it has to perform. The chip always carries the system clock, and one or more blocks of memory.

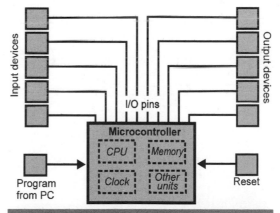

A microcontroller chip communicates with the outside world through a number of single-bit I/O pins.

The microcontroller has a number of single-bit inputs and outputs. By *single-bit,* we mean that each input or output has a single pin. We make the connection to a single pin and the input or output may be logical high or logical low. Often a pin functions as either an input or an output at any one time. It can be programmed to be one or the other. Usually we can program the pins in groups so that a group of, say, 8 pins may receive or deliver data a byte at a time.

Unlike a microprocessor, a microcontroller has internal memory. This is usually flash memory (p. 245) and is programmed by downloading from a PC. Once the program is loaded, it remains in memory indefinitely, until it is replaced by a new program. The PC uses special software which allows the user to write the program on the PC, then download it into the memory of the microcontroller. The program can then be run so that its action can be checked. If errors are found, or the user wishes to amend the program, the existing program is edited on the PC. The new version is downloaded into the microcontroller, replacing the previous version. Once the program is complete and checked, the PC is disconnected from the microcontroller, which then runs independently. Whenever the power supply to the microcontroller is switched on, it immediately starts to run the program from the beginning. There is a reset button which, when pressed makes the microcontroller run again, from the beginning.

Atmel microcontroller

The Atmel AT90S1200 microcontroller has a 20-pin dil package. It is one of the simpler chips produced by Atmel, but serves to illustrate a few more points about microcontrollers.

The pin terminals fall into four main groups:

- power supply (2.7 V to 6 V, DC).

- crystal (XTAL pins): the crystal for the system clock is connected here, the rest of the timing circuit being on the chip.

- input/output ports: there are two of these, Port B (PB) with 8 bits and Port D (PD) with 7 bits. The bits of each port can be individually programmed as input or output.

- reset pin: returns the controller to the beginning of its stored program.

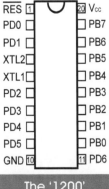

The '1200' microcontroller.

It is common to find that a pin may have two or more functions, according to the 'mode' in which it is currently running. This reduces the number of pins required and hence the size of the package. As an example, pins PB5, PB6, PB7 and the RESET pin are used for downloading the program from the PC. The program is downloaded serially, that is, as a series of single bits.

The '1200' has three kinds of memory on its chip:

- Flash memory: there is 1 kilobyte of flash memory, used for storing the program.

- EEPROM: there are 64 bytes, used for storing downloaded data.

- Data memory: 32 registers of 1 byte each used for storing data generated by a program.

The 18 I/O registers store the logic levels at the ports and also the results of certain processing operations. For example there is a bit in one of the registers to show the output from the **analogue comparator**. Two of the pins in Port B can accept analogue inputs and a comparator test to see which is the higher.

Another useful device on the chip is the **watchdog timer**. The idea behind this is rather similar to the 'time clock' used by a security guard. The guard must visit the clock at defined intervals (say, every half-hour) and press the 'OK' button. If the guard fails to do this, presumably because of some intrusion or attack, the alarm is sounded. Similarly, the watchdog timer starts a counter on the chip, and must reset this counter periodically. This is done automatically as long as the program runs correctly. It may reset the counter every millisecond, for example. However, if due to interference or noise on the system, the program does not reset the counter, it soon 'times out' and this resets the microcontroller. The microcontroller, running in the wrong part of the program, then jumps back to the beginning of the program and runs correctly from there. This is an important feature in a controller that has to run in an electrically noisy environment, such as in an automobile or on the factory floor.

PIC microcontrollers

The Microchip PICs are a large family of microcontrollers, much used in industry but also popular with hobby electronics enthusiasts. There are several hundred different PICs available, each with a selection of features that suit them to particular applications. Some of the smallest PICs are in 8-pin packages. The 12CE518 (overleaf) runs on 2.5 V to 5 V, with a frequency up to 4 MHz. It has a program memory (PROM or EPROM) of 512 bytes. It also has 25 bytes of RAM and 16 bytes of EEPROM.

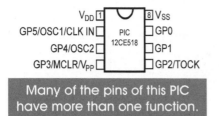

Many of the pins of this PIC have more than one function.

Programs are loaded serially and there are six I/O pins. Members of this family are RISC processors. The 12CE518 has only 33 instructions in its set, so learning to program it does not take long.

It has a built-in timer that can function as a real time clock. It also has a watchdog timer that has its own oscillator to ensure reliability.

In the early days of microprocessor development, the chips became more and more complicated as new processing facilities were added to the design. Every new processing operation needs at least one or possibly more codes to program it. A processor such as the Pentium has several hundred different codes in its instruction set. A **complex instruction set computer** (CISC) such as the Pentium is more difficult to program and the processor itself runs more slowly because of the complexities. It was found that, in practice, a relatively small number (20%) of instructions are used for the majority (80%) of operations. Consequently, a processor with a **reduced instruction set** (RISC) is simpler to build, less costly, and runs faster. Although it can not do in one step all the things that a CISC can do, it can still do them using a few steps. For these operations it may take longer than a CISC processor but, because such operations are needed only rarely, the RISC is faster overall.

Another RISC (35 instructions) microcontroller is the 16F872, which is typical of the more advanced members of the PIC family. It is in a 28-pin package, which allows it to have three I/O ports that are 6, 8 and 8 bits wide. It runs at 20 MHz. It has $2K \times 14$ words of FLASH reprogrammable program memory, 128 $\times$ 8 bytes of RAM and 64×8 bytes of EEPROM. On the same chip there are three timers, a watchdog timer with its own oscillator, a 10-bit analogue to digital converter, and a synchronous serial port. It can be programmed to capture a 16-bit value at regular intervals, which gives it applications in data acquisition. The captured data can be compared with a value in another register and produce an output signal if the two are equal. There is also a pulse width modulator to generate pulses of a set length. In total, this is a very versatile controller with many applications in control systems and measurement systems.

There is a wide range of development equipment and software to help the PIC programmer. These include assembler programs and software for programming in the C or BASIC languages.

BASIC Stamp

Microcontrollers are usually programmed by using a PC running an assembler program. However, there are other ways of programming which are easier. An example of such a way is the BASIC Stamp which, as its name implies is programmed in BASIC. The Stamp-1 is a complete microcontroller system based on the PIC16C56 and is illustrated on page 166. The PIC is seen on the right as a 20 pin IC. It has 1 Kb of one-time programmable memory into which has been written a special BASIC compiler. This has many commands that are also found in other versions of BASIC and it also has special commands applicable to the use of the PIC as a controller. The chip also has 25 bytes of ram for data storage. The 93LC56 chip seen in the photograph is a 2Kb EEPROM, which is used for storing the current program.

The Stamp-2 is an expanded version, seen in the title photograph of this chapter. Its processor is the PIC16C57, with 2Kb for storing the BASIC. The BASIC has several additional commands. The RAM is increased to 72 bytes. The chip has 28 pins, which provides for more I/O lines than in Stamp-1. In addition, the chip runs at 20 MHz compared with 4 MHz for the Stamp-1 so its action is much faster. The 24LC16B chip is a 16Kb EEPROM for programs. In both the Stamps the memory in the PIC chip is used for storing the BASIC compiler and additional memory has to be included in the system for storing the program. However, in most applications of microcontrollers, the program is stored in the on-chip memory. This may be one-time programmable for controllers programmed in quantity for use in particular applications. Other types of controller have EEPROM or flash memory that allows them to be re programmed.

There are several other microcontroller systems available one of which runs from a C compiler instead of BASIC.

Interfacing to microcontrollers is often very simple because only single-bit inputs are outputs are used in many systems. For example, input may come from a sensor such as a photodiode with a Schmitt trigger (p. 101) to give either a high or low input, depending on the light level. A switch, pressure mat, microswitch or push-button can be interfaced using only a simple pull-up resistor.

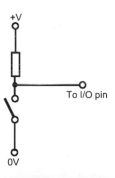

Interfacing a key switch to a microcontroller.

On the output side, the one-bit outputs of a microcontroller can switch a lamp, a motor, a siren, a solenoid, or other device through a transistor switch as on page 94. All that is needed is to wire the output pin directly to the gate of a MOSFET, or through a resistor to the base of a BJT. The transistor is switched on when the output changes from low to high. This energizes the device in the drain or collector circuit.

Microcontrollers are widely used for semi-automatic control that does not demand a controller as complex as a microprocessor. Their small size and low power requirements make them ideal for portable, battery-powered equipment. Devices in the home that may be run by a microcontroller include dishwashers, washing machines, electrical room heaters, mobile telephones, TV remote controls, security systems, and computer printers. Every day, so-called 'smart' versions of existing appliances are being designed and marketed. Automobile electronics is another fast-growing field in which microcontrollers play a significant part. Some applications of microcontrollers in industry are described in Chapter 26.

An interesting example of the increasing use of microcontrollers is found in the control system of the A320 and A330 Airbus. This aeroplane has a main control computer on the flight deck, under the control of the pilot. But the computing involved in flying the machine is not limited to the main computer. Each of the actuators that move the flight surfaces (the ailerons, slots, spoilers, elevators and rudder) has its own microcontroller. When a flight surface has to be moved from one position to another, a signal is sent from the main computer to the appropriate microcontroller in the wings or tail fins. The signal is a simple one, the equivalent of 'move the surface from angle A to angle B'. On receiving this signal, the microcontroller controls the current flowing to the actuator motor of the flight surface. Sensors report back to the microcontroller the actual position of the surface and the current to the motor is adjusted accordingly, instant by instant. The microcontroller reports back to the main computer from time to time on the position of the surface and finally reports back when the action has been successfully completed. This system is known as **distributed processing**, in contrast to central processing by a single central computer. Distributed processing reduces the amount of wiring needed in the aeroplane and reduces the workload of the main computer, leaving it free to concentrate on other perhaps more vital tasks.

23 Telecommunications

The term 'telecommunications' means 'communicating at a distance'. This could include communicating by sending a letter by post but the use of the term is restricted to electrical or electronic communications. Information may be transmitted either by electric currents (in wires) or by electromagnetic radiation. Electromagnetic radiation is usually taken to mean radio waves but visible light and infra-red are also forms of electromagnetic radiation, so communication by optical fibre is included in telecommunication. The term includes the transmission and reception of sound, vision, text and information of other kinds.

Formerly all telecommunication was analogue in nature (except perhaps telegraphic transmissions in the Morse code, which can be thought of as digital) but there is an ever-increasing trend toward digital transmission. Developments in electronics have made possible the worldwide boom in telecommunications, enabling information to be sent instantly to all parts of a country and to all parts of the world. Electronics has made telecommunications an important factor in all our lives.

Telegraphy

This is the most primitive form of telecommunications, requiring only a pair of wires, two switches (keys), two lamps or buzzers and a power supply.

Telegraphy was formerly used to send messages over distances of a few kilometres. The Morse code, with its system of dots and dashes to represent letters and numerals, was adopted as the internationally recognized system for transmitting information. It has been almost completely superseded by voice communication (telephony), although the basic idea of telegraphy still persists in the domestic door-alert and similar systems.

Telephony

Originally, this was a wholly analogue system. The telephone hand-set incorporates a microphone and a loudspeaker. Analogue signals are sent along a pair of wires to the local exchange. All the instruments in the area are connected to this exchange and here it is possible for connections between any two pairs of instruments to be made. The original manual switchboard was first replaced by a system of mechanical switches which were controlled remotely by pulses 'dialled' by the person making the call. Dialling '5', for example, sent 5 pulses to the exchange and the mechanical switch was stepped on 5 steps to position 5. Nowadays, the switching is done by logical means, using digital circuitry. The telephone instrument includes an oscillator circuit capable of producing a range of tones. The oscillator is an IC driven by a 1 MHz clock, which generates digital pulses at eight different frequencies (tones). The IC is controlled by a keyboard on the telephone and, whenever a number key is pressed, it produces two of these tones. For example, pressing key '5' produces tones at 770 Hz and 1477 Hz. The dual signal is sent to the exchange, where there is a circuit to detect which frequencies are present. In this example the circuit would identify the two tones as 770 Hz and 1477 Hz and register the fact that a '5' has been dialled. Generating and recognizing tones takes a fraction of a second and the response of the electronic switching circuits is virtually instantaneous, so that connecting a call is very much quicker than it used to be.

The principle of linking all telephones in an area to a central exchange is extended to linking all exchanges in a country or part of a country to major exchanges by long-distance line. Ultimately a world wide network of exchanges is built up.

Although a wired telephone system may seem relatively easy to operate, there are problems to be overcome. One is that the resistance of telephone wires depends on their length, so the power of the signal falls off with distance. For long-distance communication it is necessary to install amplifiers, or repeaters, at intervals along the line. There is also the problem of line impedance. When a pair of conductors are laid side-by-side there is capacitance between them. In addition, the wires have self-inductance.

The effects of resistance, capacitance and inductance amount to an appreciable **line impedance**. The mathematics of line transmission is too complicated to be dealt with here but one of its consequences is easily understood. It was stated on page 170 that for maximum transfer of power between one part of a circuit and another the output impedance of one part must be equal to the input impedance of the other. The same applies to the parts of a telephone circuit, including the impedances of the connecting lines themselves. Impedances must be matched if maximum transfer is to be achieved. This is essential to ensure that the signal reaches its destination loud enough to be heard, but this is not the only reason. If maximum power is not transferred from one part to another, some of the power must go elsewhere. If there is an impedance mismatch, part of the signal may be reflected back along the line, causing an echo. This causes serious degradation of analogue voice signals and can wreak havoc with digital signals.

With the advent of digital circuitry the telephone lines have been put into use for other forms of data transmission. Communication, particularly long-distance communication, between exchanges is often purely digital and may involve fibre-optics (see below). By using digital methods it is possible to compact the data and to transmit several sets of data at once along the same lines, and at high speed. Only in this way is it possible to maintain today's vast worldwide traffic of information, culminating in the Internet, and especially the World Wide Web. Nowadays it is easy for a person sitting at a computer in an office in, say, New Zealand to communicate with a computer in, say, Edinburgh to access instantly data stored in that computer's memory and to run software held in that computer. The time has arrived when anyone with a computer and telephone modem is able to access information from anywhere in the World.

Although national and international data transmission is becoming increasingly digital, connections to the local exchange still make use of the existing lines intended for analogue signals. The technique for transmitting digital data along these lines is usually that known as **frequency shift keying** (or FSK). This is the system adopted when a computer uses a modem to send and receive data. The computer produces data in digital form and usually sends it byte by byte to the modem along eight parallel lines. The modem has an IC known as a universal asynchronous receiver/transmitter (or UART) which converts the data from parallel to serial form. That is to say, it transmits the individual bits of a byte along a single line, one after another. It also adds several extra bits, such as a start bit, to indicate the beginning of a sequence of bits from the byte, a parity bit, and one or two stop bits to indicate the end of the sequence.

Parity

This is a way of detecting errors in data transmissions. The parity bit may be '0' or '1', and its value is selected automatically so as to make the total number of '1's sent an even number. On reception, the number of '1's received is checked and, if it is found to be odd, an error must have occurred. This system is called **even parity**. If preferred, a UART can operate under odd parity in which the parity bit makes the number of '1's odd.

The sequence of '0's and '1's is sent to the telephone line by the modem IC as a sequence of two tones. There are various systems, but typically a '0' is sent as a burst of tone at 1070 Hz. A '1' is sent as a 1270 Hz tone-burst. These two tones are generated in the modem from a single 1170 Hz tone, which is shifted up or down by 100 Hz to obtain the two tones required. This is why this technique is called **frequency shifting**. At the receiving end, the frequencies are detected by the receiving modem, which produces the corresponding logical '0's and '1's. A UART takes this stream of bits and re-converts it to bytes, removing the stop and start bits and checking the parity bit as it does so. The bytes then go to the receiving computer. Simultaneous two-way communication between the two modems along the same telephone lines is accomplished by using a second pair of tones. The modem originating the call uses 1070 Hz and 1270 Hz, as described above, and the answering modem uses 2025 Hz and 2225 Hz.

Facsimile

Fax machines (see title photograph of this chapter) are used to transmit exact copies (or facsimiles) of documents over the telephone lines. The documents may be hand-written if preferred and it may include drawings and photographs, which greatly adds to the usefulness of this method of communication. A fax machine is really a microcomputer dedicated to perform certain tasks automatically. It includes a dialling pad and circuitry, and usually a memory for telephone numbers for one-touch dialling of frequently called numbers. It has a roller which moves the document past a photo-sensitive line scanner. This produces a sequence of digital signals which is put on to the telephone line by a built-in modem. The receiving machine prints out the document using a thermal printer.

If a computer has a fax card installed and is connected through a modem to the telephone network, faxes may be sent and received directly by a computer. The fax is composed just like any other computer document, using a word-processor or graphics program. Sending the document to the fax card or modem is almost the same as sending it to a printer. Received faxes may be viewed immediately on the monitor screen. They may be saved to disk or printed out.

As telephone charges continue to fall because of improved technology and postal charges continue to rise, faxes become relatively cheaper. Further, fax transmission has the advantage of being virtually instantaneous.

The Internet

One of the major contributions of electronics toward communication and the dissemination of information has been the development of the computer Internet. Most people have heard of 'The Web' and the rapid development of e-commerce. The exchanging of e-mails has become the preferred method for business correspondence and (for many people) for private notes and news too. The subject is so vast and varied that it is not possible to deal with it adequately here. In any case, the use of the Internet is growing and developing at such a high rate that whatever we would write here is doomed to be well out of date by the time the book reaches the reader. However, it must be at least mentioned as probably the most significant aspect of telecommunications today.

Optical fibre

In the figure, a beam of light (arrowed dashed line) is passing through a transparent material (such as glass). When it reaches the boundary between the glass and the surrounding air, one of two things may happen to it. If the angle at which it meets the boundary is sufficiently small (as measured between the beam and a line perpendicular to the surface), the beam passes out of the glass into the air; it is bent (or **refracted**).

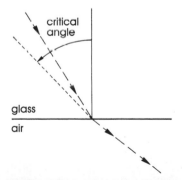

Light is refracted as it passes from a dense medium into a less dense medium such as air.

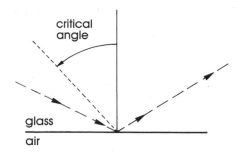

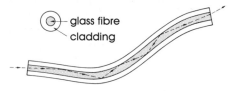

Optical fibre seen in cross-section and in lengthways section.

However, if the angle exceeds a certain amount (the **critical angle**) the beam is reflected. This is called **total internal reflection** because none of the beam passes out of the glass.

Given a narrow fibre of glass, a beam directed into the end of the fibre is reflected repeatedly until it finally emerges from the other end of the fibre. Provided the glass is of high quality and maximum transparency, very little light is lost. As shown in the diagram, the fibre does not have to be perfectly straight. Curves in the fibre make no difference to the transmission of light.

In a practical fibre-optic cable the glass fibre is about 0.1 mm in diameter. Usually it is surrounded by cladding, consisting of a layer of glass or plastic.

The optical properties of the cladding are such that total internal reflection occurs at the boundary between the fibre and the cladding. The cladding helps protect the fibre from surface scratches which would lead to loss of light. Several such jacketed fibres can be surrounded by a plastic sheath to form a multi-core cable only a few millimetres in diameter, each fibre being able to carry a separate signal. At the transmitting end of the fibre is an LED or laser diode which produces the light. At the other end is a photodiode. If the link is a long one, there are repeaters to detect a weak signal and relay it as a strong one. Since signalling is digital, there is no loss of quality on repetition.

In practice, each fibre carries several signals simultaneously, each having its own frequency. It is like having several radio stations broadcasting simultaneously at different frequencies (see later). At the receiving end each signal can be separated out (in a similar way to that in which we tune a radio receiver to just one station) and processed independently. Because of the high frequency of light, it is possible to carry many more simultaneous transmissions than can be carried on copper-wire cables.

The ability to carry a large number of channels is one of the main reasons for the adoption of fibre-optics for telecommunications. Another reason is that the cables are light in weight and inexpensive compared with copper cables. Unlike copper cables, they are unaffected by electromagnetic interference and consequently they are more secure. It is possible to use surveillance equipment to detect and read signals passing through a copper cable, but this is not feasible with optical fibre. For these reasons, optical fibre has been extensively installed on the main national and international trunk routes of the telephone system.

Radio

Radio or wireless transmission is the alternative to telecommunication by wire or optical fibre. Radio is a form of electromagnetic radiation, and like all electromagnetic radiation, is produced when atomic particles oscillate at high frequencies. It is the *acceleration* of the electrons which produces the radiation, which means that radio waves are generated by a lightning strike or when we produce a spark by opening a switch. Some of the earliest transmitters relied on the production of sparks to generate the signals. In modern radio transistors, we accelerate the electrons to move to and fro in a conductor, the aerial or antenna.

In a radio transmitter, the radio frequency oscillator is an oscillator circuit designed to produce sine waves at high frequency, most often in the range 300 kHz to 3 GHz. The frequency of oscillation may be determined by the values of the capacitor and inductor in a resonant circuit (p. 177). Alternatively, a crystal oscillator (p. 175) is used.

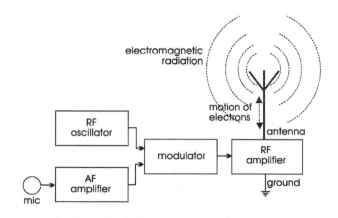

The main parts of a radio transmitter.

Given the frequency, we calculate the wavelength of the electromagnetic radiation from this equation:

$$\text{wavelength} = \frac{300\ 000\ 000}{\text{frequency}}$$

The wavelength is in metres and the frequency is in hertz. The number 300 000 000 is the approximate velocity of light (and other electromagnetic radiation, including radio waves) in metres per second.

Frequency and wavelength			
Description	Frequency	Wavelength	Application
Audio frequency*	30 Hz to 20 kHz	11 m to 17 mm	Audible sound
Very low frequency radio	30 Hz to 30 kHz	10000 km to 10 km	Experimental telecomms
Low frequency radio	30 kHz to 300 kHz	10 km to 1 km	Telecomms
Medium frequency radio	300 kHz to 3 MHz	1000 m to 100 m	Telemetry, control telecomms
High frequency radio	3 MHz to 30 MHz	100 m to 10 m	Telecomms
Very high frequency radio	30 MHz to 300 MHz	10 m to 1 m	Telecomms and control
Ultra-high frequency radio	300 MHz to 3 GHz	1000 mm to 100 mm	Television, telecomms, navigation
Super-high frequency radio	3 GHz to 30 GHz	100 mm to 10 mm	Radar, microwaves, telecomms and other devices

* vibrations in air

The output of the oscillator may be fed directly to the aerial and a signal sent by switching the oscillator on and off. By this means we could send a message by transmitting the dots and dashes of the Morse code.

Normally the radio-frequency oscillator is connected to a circuit known as a **modulator**. Instead of turning the oscillator on and off, we keep it running all the time but modulate its output. One way of modulating the signal is shown in the figure below. An audio signal is detected by a microphone (or obtained from a tape or disc player) and amplified.

The amplified signal is sent to the modulator which uses it to vary the amplitude of the radio-frequency signal. Its frequency varies with the pitch of the sound but is always much less than that of the radio-frequency signal. After modulation, the signal has its original radio *frequency* but its *amplitude* varies according to the audio signal. This is referred to as **amplitude modulation**, or **AM**.

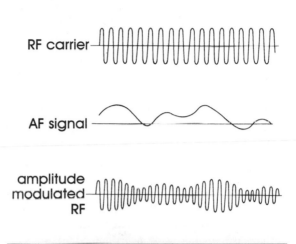

The RF carrier signal from the RF oscillator is combined with the AF signal to produce an RF signal that is modulated in amplitude at audio frequency.

Another method of modulation is **frequency modulation**, or **FM**. This has virtually replaced AM as the method used for national and regional broadcasting. This uses a different kind of modulator circuit which, instead of modulating the amplitude of the signal, modulates its frequency.

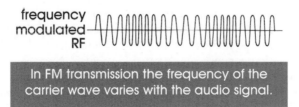

In FM transmission the frequency of the carrier wave varies with the audio signal.

The frequency of the modulated signal retains its *average* value produced by the RF oscillator but is sometimes slightly higher and sometimes slightly lower, depending on the audio signal. FM works well only when the radio frequency is high.

The form of the transmitter antenna (or aerial) is mainly dependent on the frequency being transmitted. For the lowest frequencies, the aerial may be a length of wire suspended above the ground on insulated supports. The **long wire** aerial is connected to the output of the modulator. The 0 V or ground line of the transmitter is earthed by wiring it to a metal spike embedded in the soil. Electromagnetic radiation spreads in all directions from the long wire aerial. Much of it passes upward into the higher regions of the Earth's atmosphere. An active layer in the atmosphere, known as the ionosphere, reflects the radiation several times and eventually it returns to the surface many hundreds or thousands of kilometres from the transmitter. For this reason, low-frequency and medium-frequency transmission is best for direct long-distance telecommunications, as opposed to indirect communication, using satellites (see p. 290). It is however subject to fading if the ionosphere becomes disturbed or weakened, often as the result of variations in sunspot activity.

For VHF and UHF transmission, the most favoured antenna is a **dipole**, similar to the receiving dipole illustrated opposite. It is connected to the output and ground line of the modulator. The propagation of the waves is strongest in the directions perpendicular to the length of the dipole elements. If the antenna is vertical, the radiation is strongest in the horizontal direction. For good reception, the receiver should be in sight of the transmitter, which is usually situated on high ground to cover the maximum area. Thus VHF and UHF stations are reliably received only locally and a network of transmitters is required to service a whole country.

Satellite telecommunications and the transmission and reception of super high frequency radio (microwaves) are described in the next chapter.

Radio reception

The essential elements of a radio receiver are shown opposite. The aerial may be a long wire or a dipole, depending upon the frequency to which the receiver is tuned. Currents produced in the aerial on the arrival of a radio signal are fed to a tunable resonant circuit consisting of an inductor and variable capacitor. A dipole may have additional elements to improve reception in fringe areas. Only the dipole is connected to the receiver the other two elements being unconnected. The lengths and spacing of the elements are related to the wavelength. Signals of the correct wavelength cause currents to flow in the elements which re-radiate electromagnetic waves. The result is that signals reaching the aerial from the direction of the director are received more strongly than those from other directions.

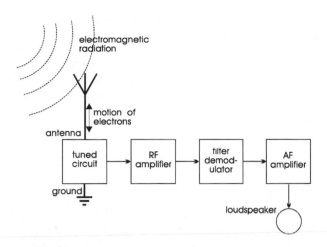

The main parts of a radio receiver.

Domestic receivers in the low- and medium-frequency bands often have an internal aerial in the form of a ferrite rod. Ferrite has a high magnetic permeability compared with air, so the magnetic lines of force of the arriving radio signal tend to be bunched together in the rod instead of passing on either side of it. This produces an alternating field in the rod. One or more coils are wound around the rod and alternating currents are induced in these. The ferrite rod with coils around it is connected in parallel with the tuning capacitor of the radio receiver, forming a tunable resonant circuit.

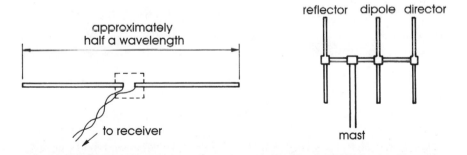

The dipole is a popular form of antenna. When it is part of an array, with one or more reflectors and directors, it becomes highly directional.

Whether the aerial is a long wire, a dipole or a coil on a ferrite rod, the result is the same. When the resonant circuit is tuned to the frequency of the carrier wave it resonates strongly in sympathy with the carrier. The amplitude of the oscillations varies according to the amplitude of the carrier which, as we have seen, is an analogue of the audio signal. This signal is amplified by a radio-frequency amplifier. It then goes to a rectifying **detector** circuit which, in a simple radio receiver, may consist of nothing more than a diode.

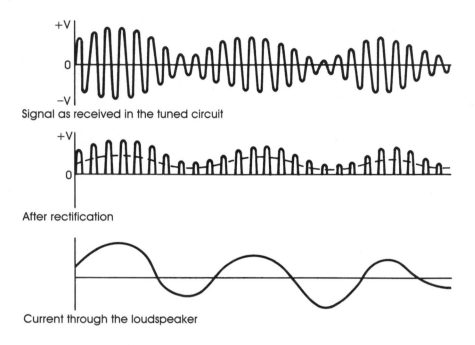

Signal as received in the tuned circuit

After rectification

Current through the loudspeaker

> Several stages are needed to extract the audio signal from the modulated carrier.

At this point, the *average* value of the signal has the same form as the audio signal, but the signal still contains its high-frequency carrier wave component. The next stage is a low-pass filter, which removes the high-frequency (that is to say, the radio-frequency) component from the signal. We say that the signal is now **demodulated.** This leaves only the audio-frequency component, equivalent to the average voltage. This is fed to an audio-frequency amplifier which increases its amplitude and makes it alternate about zero volts. The amplified signal is sent to a loudspeaker and the original broadcast sound is heard.

In an FM receiver, demodulation consists in comparing the incoming signal with a signal generated in the set and having the carrier frequency. The demodulator produces a signal dependent upon the difference between the frequency of the received signal and the carrier frequency at each instant. This difference is the audio signal. An advantage of FM is that the demodulator is not affected by variations in the amplitude of the received signal. Variations of strength due to fading have only a limited effect. Sudden bursts of radio noise such as caused by lightning are very noticeable on AM, but have no effect on FM.

Selectivity

One of the problems of radio reception is that of picking out the transmission from one particular station from a host of other transmissions. Using a directional aerial such as a dipole with one or more director elements helps to pick out the required station and a ferrite aerial too is directional. To pick out a single station we have to rely on the selectivity of the resonant circuit. Although this in theory resonates most strongly at the frequency to which it is tuned, the circuit is able to resonate quite strongly at close frequencies. For this reason a strong local station with a frequency close to the tuned one may swamp reception of the required station. It is possible to design a circuit to make it more selective and respond more strongly to one particular frequency but, unfortunately, the more selective the tuning circuit is made, the less sensitive it becomes. Distant and weak stations are not received.

The **superheterodyne receiver** (or superhet) is a solution to this problem in AM receivers. The superhet depends for its action on the phenomenon of beats.

Beats

When two signals which are almost equal in frequency and amplitude are mixed together, the result is a signal which changes regularly in amplitude, causing beats. With two sound signals we hear a throbbing sound, the rate of throbbing or beating being equal to the difference between the two frequencies. In the figure overleaf, we show two sine waves with frequencies 200 Hz and 220 Hz. When added together, the resulting signal (thicker line) varies in amplitude, or beats, at 20 Hz.

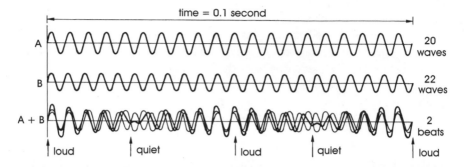

time = 0.1 second

A — 20 waves

B — 22 waves

A + B — 2 beats

loud   quiet   loud   quiet   loud

Two signals A and B, with frequencies of 200 Hz and 220 Hz, combine to give a signal (A+B), which beats at 20 Hz.

A superhet (opposite) has its own local oscillator operating at an intermediate frequency (IF). When the set is tuned to receive a station at a given frequency, the internal oscillator is tuned to a frequency that is (usually) 455 kHz above the station frequency. The two signals are mixed and this produces a new radio-frequency signal which beats at a frequency equal to the *difference* between the station frequency and the intermediate frequency. As we have already said, this difference is 455 kHz so, whatever the frequency of the station, the signal passed to the rest of the receiver is at 455 kHz. This is the *new* carrier frequency and the audio signal is still amplitude-modulated onto it, just as it was on the original carrier wave.

Now that the set has a single *known* carrier frequency to deal with, it makes subsequent processing that much simpler. First the signal is passed through a narrow band-pass filter. This cuts out signals at nearby frequencies. Then it is amplified, with the advantage that this IF amplifier may be designed to give its best performance with signals at 455 kHz. We are now at the stage of the RF amplifier in the diagram on page 273 and, from this point on, the stages of detection, demodulation and AF amplification are as before. One further modification is that part of the detected signal may be used as feedback to increase the gain of the receiver when the received signal is weak. This **automatic gain control** (AGC) compensates for differences in the strengths of different stations and for the effects of fading.

Software radio

One of the more recent developments in radio receivers is to replace the analogue circuits with their digital equivalents.

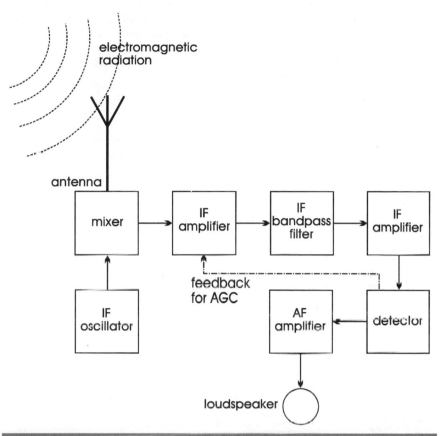

electromagnetic radiation

antenna

mixer → IF amplifier → IF bandpass filter → IF amplifier

IF oscillator

feedback for AGC

AF amplifier ← detector

loudspeaker

The main stages of a superhet radio receiver.

This has been made possible by the development of very fast analogue-to-digital converters. These are fast enough to sample and convert analogue radio-frequency signals into digital signals. The AD6644, for example, is a 14-bit ADC that can accurately convert RF signals with frequencies up to 200 MHz. This is in the VHF band and is particularly applicable to mobile telephones. All the functions such as filtering and demodulating are performed digitally, using a microprocessor. An analogue receiver relies on the precision and stability of capacitors and resistors to keep the circuits 'in tune', but digital receivers are free of this requirement. Analogue systems need adjustment during manufacture and possibly afterward. Digital receivers are programmed once and for all to operate as required. However, should broadcasting standards be changed, the receiver can be reprogrammed to meet the new standard. When standards change, analogue receivers need new circuitry and may need to be replaced. Digital radio sets can be smaller and cheaper than analogue sets.

Digital radio receivers are not to be confused with **digital audio broadcasting** (DAB) which is a complete transmisson and receiving system intended for domestic and in-car use. The audio quality of the system is equal to that of compact discs and it is less prone to interference. The analogue signal from the microphone or player is digitally sampled, then encoded by a system that reduces the number of bits required. This signal is then used to modulate the carrier wave. In the receiver, the signal is digitally demodulated and finally converted back into an analogue audio signal to be fed to the loudspeaker. As with digital tape recording and CD recording (pp. 232–3), the use of digital techniques allows the signals to be processed and checked in various ways. It also allows them to be coded or encrypted so that they can be received only by a set that has been authorized to do so.

SETI

The **Search for Extra-Terrestrial Intelligence** is an interesting application of radio telecommunication. This work is being carried on at the Arecibo Radio Astronomy Observatory in Puerto Rico. The disc of the telescope has an area of about 40 acres. This is a fixed dish nestling in a crate-like basin in the mountain. To scan the heavens it relies on the rotation of the Earth, but the angle of the telescope can be adjusted slightly by moving the receiver at the focal point. Signals received by the telescope from nearby stars (up to 10 light years away) are analysed by computer and divided into 56 million narrow-frequency channels in the microwave range. These channels are monitored automatically and the observers are alerted if anything of interest is received. The aim of the project is to detect a narrow-band signal from somewhere in Space, which would imply that it was not a naturally generated signal but that some intelligent being has sent it. So far, nothing has been reported.

One of the problems is noise (p. 189), mainly that coming from numerous sources on Earth. This effect is minimized by simultaneously scanning a star using two telescopes, the Arecibo dish and the Lovell Telescope at Jodrell Bank in England. It is unlikely that a noise source would affect both telescopes equally. Moreover, if a possible source is detected, the aim of both telescopes can be deflected slightly. The signal should disappear, then reappear when the aim of the telescopes is restored.

The receiver has extremely high sensitivity, being able to pick up signals from the spacecraft *Pioneer 10,* which is transmitting with a power of less than 10 W and is 60 000 000 000 km from Earth. The telescope also has a powerful radar transmitter used for investigating the surface features of asteroids and similar objects in Space.

Television

Like the cinema, television makes use of the persistence of vision. When light rays enter the eye and strike the retina at the back of the eye, the sensation they cause does not cease immediately but remains for some time. The sensation persists long enough for separate, motionless pictures following one another in rapid succession to give the impression of continuous motion. In television, a succession of still pictures is photographed, transmitted and received at the rate of 25 pictures a second. To the observer, the picture on the screen of the TV set appears to be moving in a life-like fashion.

The scene before a TV camera usually contains very many colours, and the same applies to the image that finally appears on the TV receiver screen. Fortunately it is not necessary to transmit details of all these colours. This is because, as far as the human eye is concerned, all colours can be made up by combining lights of three primary colours in differing proportions. The three primary lights are red, green and blue. Red and green combined in roughly equal amounts look to the eye exactly the same as yellow light, even though no light of the yellow wavelengths is present. Red and blue give purple; blue and green give cyan or turquoise. Combinations of all three produce a wide range of subtle colours. Combining all three in equal amounts produces white. All that is necessary for a full-colour TV picture is to transmit the relative amounts of the three primaries in different parts of the picture.

Before transmission, each still picture (or frame) must be converted to a series of electrical signals. Typically, a TV camera consists of a lens, a prism and filter sytem to split the focused beam into its red, blue and green components, and a CCD (p. 135) at the focal plane of each beam. Each CCD has a rectangular array of pixel units arranged in 485 rows, each with 756 pixels. The diagonal of the array measures about 15 mm. This is for the standard 525-line system; a 1920 × 1036 CCD is used for high definition TV. The digital signal from the CCD can be processed for transmitting the image by digital radio. However, the majority of domestic TV sets are capable of receiving only analogue signals at present, so the digital signal has to be processed into analogue form before transmission.

The information to be transmitted includes the following:

- Synchronizing pulse at the beginning of each frame.
- Synchronizing pulse at the beginning of each line scan.
- **Luminance signal**: the overall brightness as perceived by the eye.
- **Chroma signal** (chrominance) gives colour information.

A complex circuit known as an encoder modulates this and other information on to the carrier wave. Amplitude modulation is used. The sound signal is transmitted separately and is frequency-modulated.

Reception

The receiving aerial is a dipole with a reflecting grid behind it and numerous director elements. The processing of the signals follows the usual sequence for radio reception with the additional feature of detecting synchronizing pulses and using these to keep the scanning in the receiver in step with the scanning in the camera. The tube of a colour TV operates in a way similar to the cathode ray tube but has three electron guns, one for each colour. The screen bears dots of three different phosphors, for red, for green and blue. In the tube, immediately behind the screen is a perforated metal aperture plate or shadow-mask. It is arranged so that the three converging electron beams passing through an aperture spread out on the other side of the plate and fall on their respective phosphor dots. From the normal viewing distance the eye can not distinguish the individual dots, so the light coming from them appears to be merged. Depending on the relative intensities of the three electron beams the screen appears to have a particular colour at that point. For example, if the red and green beams are strong and the blue beam is off, the screen appears to be yellow at that point.

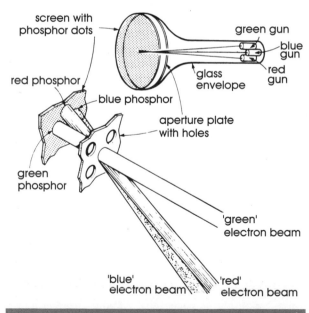

The TV image consists of dots of three colours.

Video recording

The video and sound signals may be recorded on magnetic tape in the same way as in an analogue tape recorder. The main difference is that video signals include frequencies up to 5 MHz so a very high tape speed is required. Given an ordinary audio tape-head, the tape speed would need to be over 60 metres per second. Other methods of recording and reading the tape have been devised, the most successful and popular one being the VHS system known as helical scan. The tape is wide and is wrapped round a drum as it passes from reel to reel in its cassette. In a two-head machine, the drum has two record/playback heads spaced 180° apart and the tape is wrapped around half of the drum. The tape is wrapped slantwise so that, as the drum is rotated, one of the heads traces a slanting path across the tape. As this head reaches the end of its path, the other head reaches the beginning of its path and traces the next track. The result is a closely-packed series of diagonal video tracks set side by side along the tape. In this way a greater length of track is obtained, so that the effective tape speed is increased. Additional heads of the deck lay down a control track with timing pulses and a track for the sound. In a stereo recording, there are two narrower sound tracks or, in more recent recordings, there may be more sound tracks to give a 'surround sound' effect.

Video tape recording is in the process of being replaced by digital recording. **Digital versatile disks,** usually known as **DVDs** are recorded using techniques similar to those used for recording audio CDs.

Global positioning system

Using a GPS receiver, a person can find their position on the Earth's surface and their distance above it with a precision of up to 2 metres. The system was originally developed for military and marine use but, with the inevitable fall in the production costs of the two ICs in the set, GPS receivers are now being used by bushwalkers, boating enthusiasts, and others.

The system consists of 21 satellites orbiting the Earth at an altitude of 20 200 km. Each satellite completes its orbit in 12 hours. At any time at least four satellites are above the horizon in any part of the world. The satellites transmit on only two frequencies, so that several share the same frequency. However, the digital signal from a satellite contains a pseudo-random number and different satellites are generating these numbers according to a different sequence. The GPS receiver also generates these sequences so that it can match the numbers it receives against its own numbers and thus identify the satellite.

Each satellite carries a high precision atomic clock and the transmissions from the satellites also include codes that state the exact time at which the code was transmitted. From these codes, the GPS receiver is able to calculate how long the transmission has taken to reach it. The delays tell the receiver the distance away of each satellite currently in view. The receiver has also downloaded an almanac from one of the satellites, which tells it the location of each satellite relative to the Earth. Next follows a complex routine, using this data, for calculating the position of the receiver. Distances from a minimum of four satellites are enough to locate the latitude, longitude, and altitude of the receiver.

The receiver, which need be no larger than a hand-held TV controller, is programmed to perform a wide range of tasks. Not only can it display the current position on its LCD display, but it can store the results of several position fixes (taken about once a second) as the user travels along a track, and then calculate and display the direction in which the user is moving, and the average speed. The user can download their intended route and also a set of maps into the receiver's memory. The receiver then displays the map and, on it, the location of the receiver. It can point the user in the direction and distance of the next waymark on the route and also calculate the expected time of arrival.

The satellite transmitters have a power of about 50 W. The receiver has a **patch antenna**, which consists of two rectangles of copper foil sandwiching a sheet of insulating material. It is directional, receiving signals most strongly from above. The antenna is only a fraction of a wavelength across — approximately the size of a postage stamp — and is enclosed in the case of the instrument. Yet it is able to receive the transmissions clearly, except when under trees or in a building.

24 Microwaves

Microwaves are electromagnetic waves with frequencies ranging from about 3 GHz up to about 300 GHz. The higher the frequency, the shorter the wavelength so that, at the top end of the microwave range, the wavelength is only a few millimetres. This has important effects on the way microwaves behave, as will be seen later. The range of wavelengths of microwaves puts them between the electromagnetic radiation used for broadcasting radio and TV programs and the waves of infra-red. This is why microwaves are similar to radio waves in some respects but also show some of the features of infra-red.

Generation of microwaves

Electromagnetic waves are generated by the rapid to-and-fro movement of charged particles, typically of electrons. The frequency of the radiation depends on the rate at which the electrons oscillate. Generating microwaves is just a matter of building an oscillator with the required high frequency. Oscillators of the kinds used in ordinary radio transmitters are not able to oscillate at microwave frequencies with sufficiently high power output. This is mainly because the dimensions of many of the components are greater than the wavelengths. For example, tiny capacitances that have no effect at broadcast radio frequencies exert an unduly high influence at microwave frequencies.

The whole concept of oscillator and transmitter design has to be thought out again, particularly, the kinds of components used. Some of the special techniques used for microwaves are described later in this chapter, but first we look at a well-established microwave generator, the **magnetron**. The main part of this is the anode, which consists of a copper block which has a large central cylindrical cavity surrounded by (usually) eight smaller cylindrical cavities. Running down the central cavity is the cathode, made of metal and heated by an internal coil. The whole magnetron is enclosed and its interior is a vacuum.

A strong permanent magnet outside the cavity produces a magnetic field directed along the axis of the magnetron. In operation, the cathode is made negative and the anode made positive so that there is a radial electrical field (from anode to cathode) in the cavity of the magnetron. We thus have the situation known as a **cross-field** because, at all points, the magnetic field is at right angles to the electrical field.

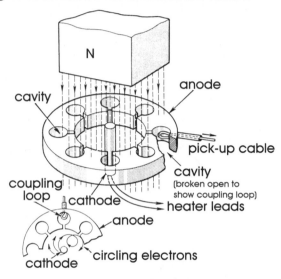

A magnetron microwave oscillator.

The hot cathode emits electrons from its surface, and these are repelled by the negative charge on the cathode. The electrons are accelerated radially toward the anode by the electric field but, at the same time, they are accelerated by the magnetic field in a direction at right angles to the magnetic field. The relative strengths of the fields are such that they describe circular paths, gradually moving toward the anode.

The eight outer cavities function as resonating circuits. A resonating circuit can be constructed from an inductor coil and a capacitor (p. 65). This resonates at a fixed frequency, depending on the value of L and C. However, there is no need to have an actual inductor in the circuit. As indicated in the upper diagram opposite, a plain wire connection has inductance. This inductance may be small at low frequencies, but increases considerably at very high frequencies. In the magnetron, the inductance is reduced to that of a single turn of a coil (the wall of the cavity), and the capacitance to a narrow gap.

Each cavity of a magnetron is equivalent to an inductor-capacitor resonant network.

By machining the cavity to the correct dimensions we produce a cavity that is tuned to oscillate at any required microwave frequency. Initially, owing to the random distribution of electrons around the cavities, a weak oscillation at this frequency occurs, causing an electromagnetic field across the aperture of the cavity. The fields of adjacent cavities are in opposite directions at any instant (they are 180° out of phase), but all fields are reversing in direction at the microwave frequency. As electrons from the cathode come nearer to the anode they are affected by these electromagnetic fields, being alternately accelerated and decelerated by the weak fields. An accelerating electron gains its energy from the field between cathode and anode, but a decelerating electron transfers its energy to the electromagnetic field of the cavity. Eventually a cloud of electrons moves around the cavity close to the anode; because of the alternate acceleration and deceleration, the cloud becomes bunched. Its decelerating regions come opposite the apertures just at the right moment to transfer energy to the electromagnetic field and strengthen it. As a result, oscillations at the microwave frequency gradually increase in amplitude and the fields become stronger. In this way, the energy from the cathode-anode field is converted to electromagnetic energy at microwave frequency. Electrons gradually lose energy in this process and, after circulating the cavity a few times, eventually become absorbed by the anode.

The electromagnetic energy, which from now on we may simply call **microwaves,** is taken from one of the cavities by means of a wire loop inserted in the cavity. Next, the microwaves may be amplified by a **cross-field amplifier**, which is very similar in construction to a magnetron. The main difference is that two metal strips (known as straps) run round the top end of the anode, connecting the walls between the cavities in alternate pairs (see overleaf). These ensure that the oscillations in adjacent cavities stay exactly 180° out of phase.

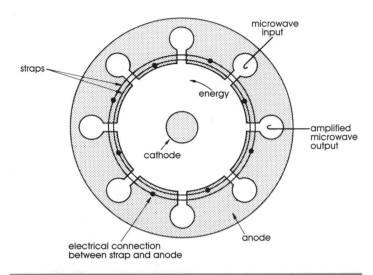

A cross-field microwave amplifier.

Microwaves from a magnetron are introduced into one of the cavities by means of a wire or loop connected to a coaxial cable (like the cable used to connect a TV set to its aerial). Action very similar to that of the magnetron causes the amplitude of the oscillation to increase and the amplified microwaves are extracted from one of the other cavities by another coaxial wire or loop.

Waveguides

Microwaves may be conducted by coaxial cables, which are convenient for many purposes, but they are more often conducted by rather more unusual means, such as waveguides. A waveguide is a metal tube, usually rectangular in section. We can think of the waves as passing along inside the tube, being reflected off the walls of the tube, zig-zagging from one end of the tube to the other. This is similar to the way in which light is conveyed along optical fibre.

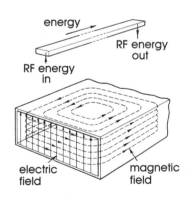

When a waveguide is conducting microwaves, the electric field is perpendicular to the guide walls and the magnetic field is tangential to the guide walls.

The advantage is that the microwave is confined to the interior of the waveguide. It does not radiate from it and cause interference with surrounding equipment, and there is very little loss of energy, so transmission is very efficient, especially at the highest frequencies. Note that the waves are passed along the waveguide in the air-filled cavity, not conducted in the metal walls, which are there solely to reflect the waves. Depending on the dimensions of the waveguide and the wavelength of the microwaves, there are various modes of transmission. One mode commonly used is the TE10 mode shown in the lower diagram opposite, in which the guide is exactly half a wavelength wide.

Microwaves are introduced into a waveguide by inserting a coaxial cable in a cavity of the magnetron or amplifier and running it into one end of the guide. A similar cable at the other end of the guide may be used for collecting the microwaves.

Microstrip

Microstrip is a technique for microwave circuit construction. The circuit is formed by a metal (often copper or gold) conducting strip on a non-conducting substrate such as alumina. The conductors are formed on one side of the board, the other side of the board having a continuous film of metal to produce a ground plane. The strips may be formed by etching the patterns on one metallized surface of the board, in a manner similar to that used for producing PCBs. Another technique is to screen-print the circuit on uncoated board, using thick conductive ink. Microstrip circuits can also be fabricated as integrated circuits using gallium arsenide.

The diagram shows the electrical field when a microwave signal is passing along a microstrip transmission line. The fact that the dimensions of the strips (a few millimetres wide) are of the same order as the wavelength of the microwaves gives rise to effects which have useful applications. Instead of thinking of the strip as an ordinary conductor, we refer to it as a **transmission line**.

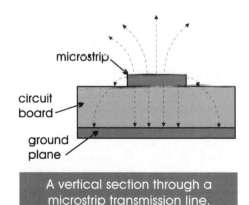

A vertical section through a microstrip transmission line, showing the electrical field.

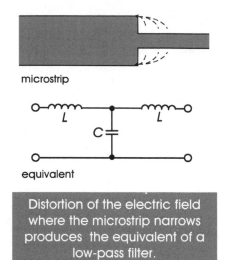

microstrip

equivalent

Distortion of the electric field where the microstrip narrows produces the equivalent of a low-pass filter.

If the width of the conductor changes suddenly at a given point, the electromagnetic field is distorted. The effect is the same as if there was an inductor-capacitor network at that point. In other words, simply by varying the width of the strip we are able to 'construct' passive components on the circuit board. On the left, we have the equivalent of a low-pass filter.

This technique can be applied to create the equivalents of other components such as transformers and mixers.

Other effects are obtained by using parallel strips with a gap between them, the coupling of fields between the strips having the effect of an inductor and capacitor in series. The circuits are 'tuned' to any required frequency by their widths, the lengths of each section and the spacing between adjacent sections.

Microstrip is used in conjunction with other special devices suited to high-frequency operation. The **varactor diode** is a microwave version of the varicap diode, and is often used as a tuning capacitor in transmitters and receivers. Another technique for tuning depends on using a sphere of a ceramic material. Essentially, the material consists of yttrium and iron oxides formed into a garnet material and are therefore known as **YIGs**. The sphere, which is between 0.2 mm and 2 mm in diameter, is placed close to a narrow loop of microstrip, and is exposed to a steady magnetic field, produced by an electromagnet. When microwaves pass along the strip they induce electromagnetic vibrations in the YIG at the same frequency. At a fixed frequency (depending on the strength of the applied magnetic field) a resonating field is set up in the YIG. This induces the microwave circuit to oscillate at the same frequency. The resonant frequency is set by the strength of the field produced by the electromagnet. By adjusting its field strength electrically we are able to tune the microstrip circuit to resonate at the required frequency. YIGs produce precisely-tuned frequencies.

Discs or cylinders of ceramic materials such as barium titanate and zirconium titanate are coupled to microstrip circuits to be used as resonators in a similar way to YIGs, except that they do not depend on a magnetic field for tuning.

Microwave telecommunications

A microwave telecommunications transmitter consists of a microwave generator, producing a carrier wave which is then modulated by the audio or other signal (for example, computer data). The principles of modulation and demodulation at the receiver are similar to those employed in ordinary radio, as described in the previous chapter. Frequency modulation is preferred.

Microwaves are transmitted into the atmosphere by using a dipole aerial of suitable dimensions. This type of aerial may also be used at the receiver. Dipoles, in which the antenna is approximately half a wavelength long, are suitable for microwaves of lower frequencies, such as a few hundred megahertz. Part of the radiation travels over the Earth's surface (the **ground wave**) and part is projected upward (the **sky wave**). Although a dipole works well with transmission at ordinary long wave, medium wave and short wave frequencies, the ground wave rapidly becomes attenuated at frequencies greater than about 1 MHz. Reception by ground waves is restricted to distances of a few tens of kilometres. Dipoles are used for the transmission of broadcast radio and TV signals, including transmissions to mobile telephones.

Because of the limitations of distance, numerous TV and radio stations are set up throughout the country to cover local areas. Similarly, the country is divided into cells, each with its own transmitter to communicate with mobile telephones. The sky waves from a dipole are of little use for communications because, unlike normal radio wavelengths, microwaves pass through the ionosphere and are lost into space.

We use a parabolic reflector for transmission at higher microwave frequencies. A waveguide conveys the microwaves to the focal point of the reflector. They emerge and are focused into a more-or-less parallel-sided beam. At the receiver, a similar parabolic reflector concentrates the arriving waves into another waveguide which conveys the waves to the receiving circuit. The advantages of the reflector antenna is that it allows transmission over relatively long distance with minimum loss of power. Transmission is by line of sight, so transmitter and receiver must normally be situated on high ground, with the antennae mounted on towers. Since the beam is highly directional, only those receivers in line with the beam can receive the signals. But high-frequency waves are attenuated by water droplets in the atmosphere, such as fog and rain, which further limits transmission distances. To cover long distances, it is usual to set up a chain of repeaters which receive a signal, amplify it, and then re-transmit it to the next station along the line at a slightly different frequency.

Nowadays, with the improved performance of optical fibre transmission (which does not require line of sight, and is not subject to outside interference or climatic attenuation) many microwave links are being replaced by optical fibre links.

Microwaves are of particular importance for satellite communications since transmissions at frequencies of 3 GHz or more are not reflected by the ionosphere. TV and other transmissions are directed up to a 'stationary' satellite, using a parabolic reflector to produce a narrow beam of radiation. The satellite is not actually stationary. It is in orbit above the equator, but its orbiting time is exactly 24 hours, so keeping the satellite in position above a fixed point on the Earth's surface. We say that it is in **geostationary orbit**. From its high altitude, several tens of kilometres above the Earth's surface, the satellite can be 'seen' from a little less than half the Earth's surface. It can also 'see' the same area.

The transmission is directed up to the satellite from the transmitting Earth Station and received by the satellite. It is amplified and then re-transmitted on a slightly different wavelength, to avoid interference between the incoming and outgoing signals. For telecommunications, a narrow downward beam may be used but, for TV broadcasts intended for direct reception by viewers over a country-wide area, the downward beam is wider. On the ground, the signal is received by a parabolic reflector and fed to the receiver circuit.

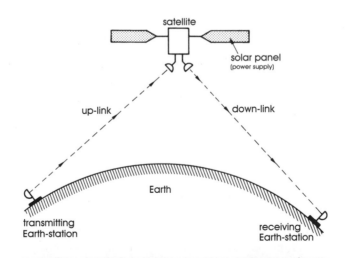

A geostationary satellite can provide microwave communication between two stations that are not in line of sight.

In addition to the main telecommunications channels, the satellite is in microwave contact with its ground station, which receives data concerning the status of the systems inside the satellite and its location in space. The ground station can also transmit instructions to the satellite, including instructions to power up its small rocket motors to correct its attitude from time to time.

With three geostationary satellites, it is possible to cover the whole of the Earth's surface, except for the polar regions. Unlike cable links, the cost of satellite links is independent of the distance. In underdeveloped areas, where reliable land lines are scarce (or non-existent) and costly to install, satellite communications bring great advantages.

Radar

Radar (or **ra**dio **d**etection **an**d **r**anging) was first demonstrated in 1935 by Sir Robert Watson-Watt. Subsequently, it became of major importance for detecting aircraft in World War II, a fact which did much to accelerate its development. The principle is simple. A transmitter sends out a series of very short pulses of microwaves. These travel outward from the antenna, being reflected off any object that they strike. Part of the radiation is reflected back to the location of the transmitter, where there is a receiver. Electronic circuitry measures the time passing between transmission of a pulse and the reception of its echo. The time taken is proportional to the distance of the object, so the distance of the object can then be calculated.

Although radar can be used solely for measuring distances, it is more often used for measuring direction as well. The antenna is made to rotate continuously on a vertical axis so that it scans the surrounding area repeatedly. One form of antenna, often used on ships for detecting other surface craft at night and in fogs, is shown overleaf. The waveguide conveys pulses of high-energy microwaves from the amplifier directly to the antenna horn. With the typical cheese antenna, the vertical height of the beam is large, but this is not of significance. The beam is narrow in the horizontal plane, making it easy to pin-point the direction of the detected object. In aircraft radar systems such as are in use at airports a huge rotating array of parabolic reflectors, similar to a small radio telescope (p. 182) directs a radar beam up into the sky.

A more advanced form of antenna is the **phased array**. This consists of an array of dipoles mounted on a flat plate. The emission from each dipole is timed so that the wavefronts are slightly out of phase. The effect of this is the same as if a single large antenna is producing a beam *at an angle* to the direction of the array (see over).

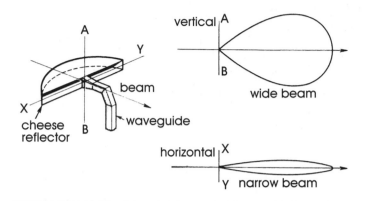

The beam of a cheese antenna is very narrow on the horizontal plane, giving good directional properties.

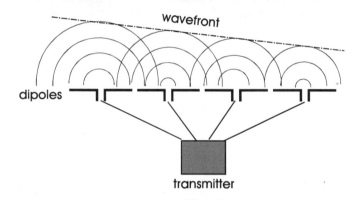

Delays in the signals going to the right-hand dipoles cause the wavefront to propagate at an angle to the plane of the dipoles.

This makes the beam electronically steerable, so that it can be swept over a wide angle without the need for a rotating antenna. The Westinghouse E-3 airborne warning and control system (AWACS) has a phased array inside its saucer-shaped radome. A similar steerable receiving system is based on the inverse of the phased array.

Most often the information obtained from the rotating antenna is displayed on a **plan position indicator**. This consists of a cathode ray tube with electrodes arranged so as to produce a narrow fluorescing line which rotates about the centre of the screen.

Its rotation is synchronized with that of the antenna so that if, for example, the antenna is directed north, the line is directed vertically up the screen. As the line is scanned from the centre to the margin of the screen, it displays reflected signals as an increase in brightness of the line.

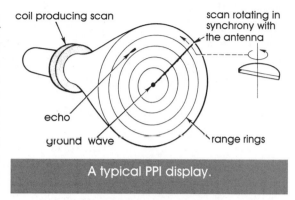

A typical PPI display.

The distance of the bright spots from the centre of the screen is proportional to the distance of the reflecting objects from the radar station. There is also a bright spot at the centre of the screen due to the reflection of the ground waves from nearby buildings and other structures.

Instead of the PPI the display may be a computer monitor. This allows for more sophisticated processing of the information and for the simultaneous display of ancillary information on the screen. For example, the computer can distinguish between stationary objects that show up on every rotation of the antenna and those which have moved since the previous scan. This may be useful in certain situations, for example to pick out the movements of vessels in a harbour. Once the distance and angular position of an object has been measured, it requires only a simple computer program to log this data regularly and then calculate the velocity of the object and also its acceleration or deceleration. The ability of radar to measure velocity is exploited in radar traps for speeding motorists.

High-resolution radar

High-resolution radar uses microwaves of high frequency and hence of the shortest wavelengths. This enables much more detail of the shape and structure of the target to be revealed. In addition, the Doppler effect (see overleaf) causes the microwave frequency to be slightly raised when radar is reflected from surfaces moving rapidly toward the receiver. The opposite occurs if the surfaces are moving away. Frequency analysis of the reflected signal detects rapid motion, such as produced by the blades of a turbine jet engine. If a computer has a library of possible target shapes (including jet engines) at high resolution, it can search through this and may be able to identify the target as an aeroplane of a given type.

Sub-surface radar

Microwaves are able to penetrate beneath the surface of the soil and reveal the presence of buried objects. Under favourable soil conditions, objects as far down as 50 m may be detected. The equipment consists of a transmitter to send pulses of wide-band microwaves into the soil and a receiver to detect radiation reflected back from buried objects. Whereas in the case of conventional radar the transmitter and receiver are stationary and the detected object is moving, with sub-surface radar the transmitter/receiver is moved in a regular pattern across the area being scanned and the objects themselves are still. The signals from the receiver are fed to a computer for analysis. This produces a display showing a vertical section through the strip of soil currently being investigated. This may be in colours that distinguish the different types of material present below the soil. If several scans are carried out along parallel grid lines, a picture of the sub-surface features of an area of terrain may be built up.

Applications of sub-surface radar include finding buried cables and pipes, locating persons buried under snow avalanches, and in archaeological surveys. One researcher using SSR has discovered a boat buried high in the mountains of Turkey; this is possibly the remains of Noah's Ark.

Doppler systems

These are used for detecting moving objects. Perhaps one of their most familiar applications is for automatically opening the doors of a shop whenever a customer approaches. The **Doppler effect** is noticed in everyday life if we are standing on the pavement when a fire-engine rushes by at high speed. As the vehicle passes, the apparent pitch of its siren abruptly falls. The frequency of the sound is higher when the fire engine is approaching, and lower when it is receding. A similar effect is found with reflected microwaves. Microwaves of a given frequency are sent from a transmitter mounted above a door. If a person is approaching the door, some of the waves are reflected back toward the transmitter and have a slightly higher frequency than the transmitted waves. The change of frequency is only very slight, but it is measurable. For example a person walking toward a transmitter at 6 km/h causes the frequency of a 10 GHz microwave beam to increase to 10.000000056 GHz. This is an increase of only 56 Hz. To detect this increase, the detector circuit mixes the received signal with part of the transmitted signal. The two signals interfere with each other to produce a beat frequency of 56 Hz (p. 275).

The beats are detected electronically and trigger a circuit that opens the door. Note that this action depends on the person being in motion. If two persons are standing talking outside the door, but are not moving, the reflected signal has the same frequency as the transmitted, there is no beating, and the door remains shut.

A similar principle is used on some car alarm systems. The transmitter/receiver is usually mounted on or near the floor of the vehicle, the microwaves being reflected by the metal floor to give a wide area of coverage. Any moving object inside the car causes a shift of frequency, which is detected and triggers the alarm to sound. It is also possible to measure the rate of beating and thus find the velocity of the moving object. This is the basis of the devices used by police to measure the speed of vehicles, using equipment positioned at the roadside.

Microwave lamps

Microwaves are invisible to the eye but they can produce visible light when a beam hits a plate coated with a mixture of indium compounds and bromides. The composition of the light is similar to daylight. The main advantage of such lamps is that they last 60 000 hours, which is far longer than the life of ordinary filament lamps. These lamps have a power of 50 W and are suitable for domestic use.

Microwave ovens

Microwave energy is readily absorbed by the atoms and molecules of any non-conducting material it passes through and is converted into heat (thermal) energy. The microwaves behave in much the same way as their nearest relatives in the electromagnetic spectrum, infra-red radiation. A typical microwave oven is featured in the title photograph of this chapter. In microwave ovens, the frequency (usually 2.45 GHz) is such that it is efficiently absorbed by water molecules, which are nearly always present in high proportions in the food. The microwaves are generated by a magnetron and conducted by a waveguide into the cooking section of the oven. The cabinet of the oven is made of metal, so it acts as a totally enclosed waveguide, confining the microwaves to the oven. A switching mechanism ensures that it is not possible for the magnetron to be activated while the door is open. Although the door has a transparent panel to allow inspection of the food while cooking, this panel is covered with a screen of perforated metal to reflect microwaves back into the interior.

To promote even cooking in all but the cheapest ovens, the food is placed on a slowly rotating turntable. There is a timer circuit to allow the oven to operate for a period up to 99 minutes, to the nearest second. Most actual cooking times are much shorter than this, usually only 5 to 10 minutes. There is an automatic switch to turn the magnetron on and off for a varying proportion of the time, to provide different heating levels. Times and heating levels are visible in a small LED display.

Microwaves penetrate food and cooking containers easily so that heating occurs more-or-less evenly throughout the food. This appreciably reduces cooking time when compared with a traditional thermal oven. It makes microwave ovens (usually with a 600 W to 1 kW magnetron) much more convenient and economical to operate than a conventional electric oven, whose heating elements may require 3 kW or more for much longer periods. In addition, the walls of a microwave oven are made of metal, so they do not absorb microwave energy. After a period of cooking, the walls are only slightly warmed by convection from the cooking food. By contrast, the walls of a conventional oven become very hot indeed, a source of accidental injury. One slight disadvantage is that cooking with microwaves does not readily produce the crusty or roasted outer layer that is familiar on cakes, roast joints and similar foods cooked in a conventional oven, but there are ways of producing this effect by other means.

When it is energized, there is a p.d. between the cathode and anode of the magnetron of around 4000 V, and it is capable of delivering up to a kilowatt of power. This is why it is essential for a microwave oven to be serviced only by a trained engineer with specialized equipment.

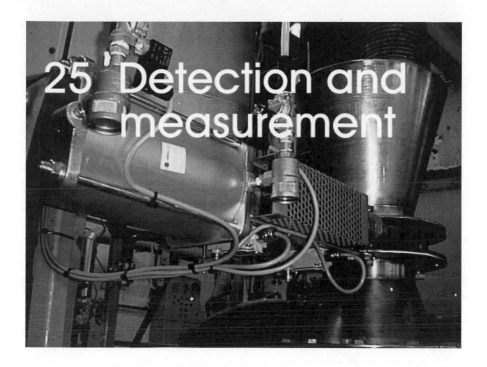

25 Detection and measurement

Detection of events and the measurement of quantities are two interrelated tasks of electronics. In this chapter we look at a range of instances of how electronics is applied to these tasks. In the following chapter we see how we may use the information so gathered.

There are several reasons why electronic methods are so often superior to other techniques for detection and measurement:

Speed: the response time of electronic sensors is usually of the order of milliseconds or microseconds.

Sensitivity: electronic circuits can detect events or make measurements that are totally out of range of human perception.

Safety: electronic devices can perform their tasks in environments where it is dangerous for humans to venture.

Remote operation: electronic devices can operate at the far end of a communications link.

Accessibility: electronic circuits are easily connected to logical or computing systems so that data may be stored, processed, or acted on in other ways.

We now examine a selection of detection and measurement applications of electronics to illustrate the points listed above.

Electronic counting

'How many?' and 'How much?' are frequently asked questions. We might want to know, for instance, how many units of a given product have been assembled by a factory machine. A method which may be used for counting objects on a conveyor belt is based on the rapid action of a photodiode. The photodiode is placed opposite a light source so that each object interrupts the light beam as it passes. Often the beam comes from an infrared source and the photodiode has a filter that admits only infra-red. In this way, counting is unaffected by changes in light levels in the factory. When an object interrupts the beam, the photodiode becomes less conductive. The drop in the amount of current passing through the photocell is detected as a fall in the p.d. across a resistor in series with the photocell. The falling p.d. triggers a digital counter. Not only can this system count the total number of objects in a given period of time, but it can also provide an output signal when a certain number of objects has been counted. The signal may be used to initiate some further action, such as switching on an indicator lamp or stopping the conveyor belt. A similar method is used to count the number of cars entering a multi-storey car park. With two counters, one placed at the entrance and another at the exit, it is possible to keep a tally of the number of cars in the park and to display warning notices and close the entrance gate whenever the park is full.

Counting manufactured items is related to the task of stock-taking. One way in which electronics helps with this is by use of **bar codes**. The detector circuit is similar to that used on a production line, though the subsequent processing is very different. Once again we are relying on the rapid action of the light sensor. A bar code is a form of information that is readable by scanning a printed array of bars. The product bearing the bar code is passed across a window and scanned by a high-intensity, narrowly-focused beam of red light, produced by a laser. In some forms of bar-code reader, a stationary hand-held reader is held against the bar coded area and scans the bar code. The reflected light is sensed and the information is translated into a series of time intervals. The bars and the spaces between them are either one, two, three or four units wide. Scan times depend on how fast the bar code is wiped across the window. Times are longer if the scan is not exactly perpendicular to the bars. But it is the *relative* lengths of the times that matter. These are determined by analysing the output from the sensor by a computer program.

Numbers are coded according to a system in which each number is seven units wide. For example, the figure 8 is coded by the seven binary digits 0110111. Printed as a bar code, this is represented by a space one unit wide, a bar two units wide, a space one unit wide and a bar three units wide.

The codes are such that each can be identified whatever the scanning direction; for instance, there is no figure coded by the reverse of the above code, 1110110. But the inverse code has a meaning; for example, the code 0110111 and its inverse code, 1001000, *both* represent the figure 8. The application of this is that most bar codes consist essentially of two numbers, each of five or six digits, one specifying the manufacturer, and the other specifying the nature of the goods. One number is coded directly and the other inversely, so it is easy for the computer to know which number is which, whatever direction the code is scanned in.

The bar code also includes a single digit indicating the nature of the goods, for example 9 for all books and magazines. A final digit is used as a check number. This is calculated from the previous numbers, according to a formula. On reading the other numbers the computer calculates what the check number should be. If this does not agree with the check number read from the bar code, it indicates that there has been an error in reading the code, and the operator is warned accordingly.

Bar-code readers enable a company to keep a complete list of stock on its computers, adding items to the stock list as they are manufactured or delivered from another company, and removing items from the stock list as they are dispatched to customers. Computers can do more than simply keep count; they can analyse the flow of goods and produce invoices, forecasts of stocks and other useful information.

Bar codes are used wherever an assortment of goods has to be listed. This includes sales receipts in stores. Almost all supermarkets and many other retail stores have a bar-code reader linked to the cash register. Usually, information such as price is not coded. This can be held on a central computer and, once the product has been identified from its bar code, the computer looks up the price and sends it to the checkout till. This allows prices to be changed in the central computer by the retailer should there be price increases or special sales offers.

Measuring flow

Measuring the flow of liquids and gases is an important aspect of industrial processing. We may need to measure flow of combustible gases to furnaces, the flow of coolants, or the flow of various reagents in chemical plant. The most commonly used technique is to use a rotating vane inside the pipe through which the fluid is passing. The rate of rotation is proportional to the rate of flow.

Various methods are used for measuring the rate of rotation. In some flow sensors, a beam of light is passed through the pipe to a light sensor placed on the opposite side. The beam is broken as each blade passes through it, producing a pulsed signal from the light sensor. This is easily converted to a measurement of the flow rate. In other flow sensors, the vane is made of stainless steel and a magnetic sensor outside the pipe detects the blades as the vane spins. A pulsed signal is generated which can then be processed to measure the rate of flow.

A rotating vane in a pipe restricts the flow to a certain extent. Measuring the flow of liquids in pipes with no restriction to the flow is achieved by the arrangement shown below.

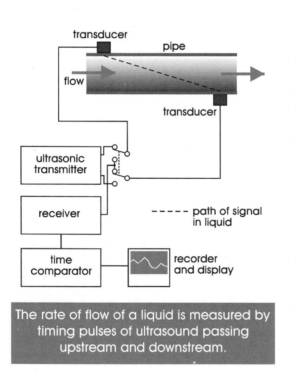

The rate of flow of a liquid is measured by timing pulses of ultrasound passing upstream and downstream.

A pair of ultrasonic transducers are mounted in the pipe as shown. An automatic switching arrangement makes one transducer act as an emitter of ultrasound while the other acts as a receiver. An instant later their functions are reversed. In this way the equipment measures the time taken for pulses of ultrasound to travel downstream and upstream. The difference between these times is directly proportional to the rate of flow. The time differences are calculated by suitable logical circuitry and the flow rate displayed or recorded.

Another technique relies on the Doppler effect, and is suitable for liquids that contain either solid particles or bubbles. In this technique the two transducers are placed side by side in the pipe directed against the flow direction. One is a transmitter and the other a receiver as in the usual arrangement for Doppler effect measurement. Ultrasound is directed toward the oncoming bubbles of particles, is reflected from them and is detected by the receiving transducer. The beat rate gives a measure of the rate of flow of the liquid.

The **venturi** is a method of measuring flow. It depends on the reduction in pressure that occurs when the flow passes though a narrower tube. This device is used for measuring wind speed and the airspeed of aircraft. A differential pressure gauge measures the reduction in pressure, which is proportional to the square of the flow rate.

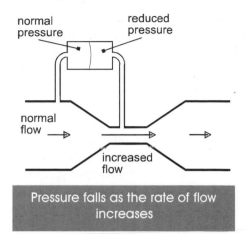

Pressure falls as the rate of flow increases

An ingenious technique for measuring flow depends on Corioli's Force. This is the force that causes a vortex when water is draining from a bath or washbasin. The flowing liquid is passed through a metal tube shaped in an almost complete circle, like the Greek letter Ω. Coriolo's force distorts the tube by an amount proportional to the rate of flow. Distortion is measured by strain gauges mounted on the outside of the tube. This method has the advantage that the liquid is totally enclosed in a metal tube. The technique is particularly suitable for flammable or highly toxic liquids, provided that they do not corrode the tube.

Electronics is also able to measure the rate of flow of gases. An ingenious gas flow sensor together with the necessary circuit can be fabricated on a silicon chip. As the gas flows through the orifice it presses against the springy strip. The faster the flow the greater the pressure and the greater the distance between the gold plates.

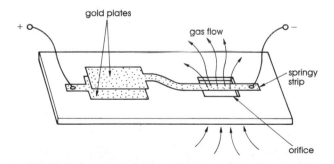

This device converts the rate of gas flow into an easily measurable variation in capacitance.

The attached circuit measures the capacitance between the gold plates and, because this depends on their distance apart, it is proportional to the gas flow. Such a device is very small and could replace the domestic gas meter. Moreover, it lends itself to remote reading, so that instead of the meter reader having to call at the house (to find the occupants away for the day) the meter is read from a central station.

Whether we are measuring the flow of liquids or measuring the flow of traffic, electronics has an application. On today's busy road systems, it is essential that the traffic should be kept moving. Electronics makes many contributions to this. One example is **Trafficmaster**®, which is a system for warning motorists of traffic jams and similar causes of delays on motorways. Infra-red sensors mounted on bridges over the motorways measure the speed of traffic passing beneath them, and the number of vehicles passing. The speeds are averaged for successive three-minute periods and, when the average falls below 30 mph, a signal is sent by radio to the control centre of the organization. Incoming signals are analysed by computer and the results are sent by radio to a pager mounted on the dashboard of the car. The information is presented to the driver in the form of a motorway map showing the points where jams are occurring, and the average vehicle speeds at these points. Sensors are two miles apart on average, so three adjacent points indicating low-speed traffic implies a tail-back at least four miles long. The driver can also call up text messages explaining the causes of the trouble in more detail. The information is updated every three minutes, giving a continuous assessment of the traffic situation. A system such as this, which works just as well at night and in thick fog, provides information aimed at increasing road safety, cutting costs and travel time, and reducing pollution.

Measuring levels

In many automated industrial processes it is essential to know the level of a liquid or powdered substance stored in a tank. The level can be measured by using **ultrasound**. Ultrasound is sound of such high frequency that it can not be heard by the ear. Frequencies of 16 kHz or higher fall into this category, though frequencies as high as 40 kHz are often used. Ultrasonic techniques have a similarity to radar. The ultrasound is generated by an oscillator running at the required frequency. A piezo-electric transducer, cut so as to resonate strongly at the required frequency, is used to convert the electrical oscillations into ultrasound waves. A similar piezo-electric device is used to detect ultrasound; this is connected to a circuit that indicates the amount of ultrasonic energy being received or the timing of reflected pulses.

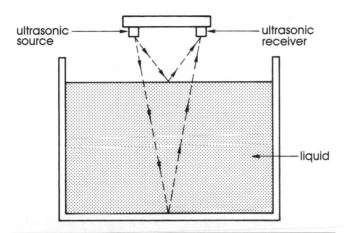

An ultrasonic method for measuring the depth of a liquid.

A source of ultrasonic pulses is directed downward on to the substance in the tank. Part of the ultrasound is reflected from the top surface of the liquid and detected by the sensor. Part is detected after reflection from the bottom of the tank. The difference between the times of arrival of the two reflected pulses is a measure of the depth of the liquid. This technique is equivalent to using a dip-stick but there is one important difference. There is no easy way in which the information obtained by using a dip-stick can be passed automatically to an electronic control circuit. By contrast, the time differences measured by the circuit of the ultrasonic equipment may be fed directly to other electronic circuits.

The same feature applies to the other measuring examples quoted in this chapter; all of them are able to pass information directly to a computer or other electronic controller.

As with radar, the technique can be extended to build up a picture of the location of surfaces and objects within the field of view of the receiver. An application in which this has been very successful is the ultrasonic scan of the human body. In ultrasonic scanning, bursts of ultrasound are applied in a narrow beam to parts of the human body, for example the skull and the abdomen. The ultrasound penetrates the tissues but part of the beam is reflected at the boundary layer between organs, where sudden changes of tissue density occur. The reflected ultrasound is detected and information is fed to a computer.

This produces a picture on the screen of its monitor, showing the internal organs of the body as they would be seen in section. The technique provides information which can be used to support or replace examination by X-rays. The ultrasonic scanner is of particular use in examinations during pregnancy. The frequent use of X-rays in such circumstances could cause genetic damage to the unborn child and to the mother herself, whereas ultrasound is relatively safe.

Measuring position

In an automatically-controlled machine, it may be important for the control circuit to 'know' the exact position of certain parts at any given instance. It may be essential to know the position of the work-piece or of a tool such as a drill bit, or that a safety gate has been shut. It is not enough to have a motor or piston that moves a part of the machine to a given position. Friction, jamming, a 'spanner in the works'or some other such mishap may have prevented the part from reaching its intended location. A positive measurement of its position is required. Some devices for position measurement have been described in Chapter 11. Here are two more.

The simplest type of position sensor, and one that is perhaps used in industry more often than other types, is the **limit switch**. This may be no more complicated than a microswitch, or it may be an optical broken-beam sensor, or a Hall-effect device to detect the presence or absence of a metal machine part. There are two limit switches in the title photograph of this chapter, located beneath the perforated safety casing. Their function is to detect the position of a sliding shutter that control the flow of reagents in the plant.

The switch on the left detects when the shutter is fully open; the switch on the right detects when it is fully closed. These are the only two positions of interest in this case. The output from the switches is a simple binary, easily translated into a '0' or a '1' for processing by the logic circuits controlling the plant. Limit switches are often used to register the position of a cutting tool where workpieces are to be machined to standard dimensions.

In other systems, it is important to know the exact position of the machine part as it travels between its limits. The system illustrated opposite is applicable to a part that slides to and fro in a straight line. A sliding panel with clear and opaque sections is attached to the sliding part. A row of four photodiodes senses infra-red radiation coming from behind the panel. In the position shown in the figure, in which all four sections are clear, infra-red reaches all the photodiodes.

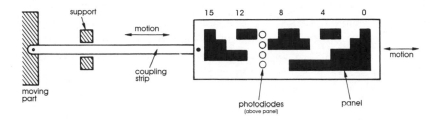

As the part and panel move from step to step, the illumination of only one diode changes.

The photodiodes are each connected to a logical circuit and, in this example, all four outputs are high (1111). The panel (and therefore the sliding part) has sixteen positions, numbered from 0 to 15. Since there are four photodiodes, the obvious thing to do would be to set out the pattern on the panel to represent the binary numbers from 0000 to 1111 in numerical order. Instead, as can be seen in the figure, the pattern gives 0000 for position 0 (as might be expected) but 1000 for position 1, 1100 for position 2 and so on. Each position is represented by a four-digit *code*, known as the **Gray code**. The point about the Gray code is that, as the panel moves from one position to the next, only one digit is affected at each step. This helps eliminate a source of error. It is not easy to be sure that the row of photodiodes is so straight and is aligned so precisely that changes in the panel position affect all four of them at exactly the same instant. It is more likely that they are affected in some random order. For example, if positions 7 and 8 were coded as ordinary binary numbers, 0111 and 1000, the change from one to the other might go through this sequence:

Digit	Step 7	... changing to ...			Step 8
A (LSB)	1	1	1	*0	0
B	1	*0	0	0	0
C	1	1	1	1	*0
D (MSB)	0	0	*1	1	1

The asterisks indicate the points at which the photodiodes are affected by the motion of the panel. In this example they change in the order BDAC, and generate numbers 5, 13 and 12 while doing so. This gives completely false readings of the position as the panel moves. In moving from positions 7 to 8 with the Gray code, the change is from 1101 to 1100; only digit D is affected. The other digits remain as before, so the order in which they are affected does not matter and there are no spurious intermediate states.

Special ICs are made for converting the Gray code into its binary equivalent. Being digital, the output of this is ideal for sending to any other logic circuit or to a computer for use in automatic control of machinery.

Detecting current

A survey of the examples quoted in this chapter shows that the final stage in an electronic detection or measurement is concerned with voltage, current, capacitance, inductance or timing. The majority of circuits depend on voltage, capacitance and timing. A few, such as the inductive loops buried in the road surface to detect vehicles approaching traffic lights, depend on induction. A few depend on current. Detection of excess current is the basis of the **earth leakage circuit breaker**. This is a safety device intended to switch off the mains supply to electrical equipment when a fault causes excess current to flow to Earth. The circumstances in which this action is most vital is when the current is flowing through the human body, with high risk of the person being electrocuted.

ELCBs in the home usually take the form of a small plastic device that plugs into a mains socket. It carries another mains socket to supply power to the protected appliance. All kinds of equipment, from electric mowers and power drills, to garden fountain pumps immersed in a pool, should be protected by an ELCB. The other form of the device is slightly larger and is mounted in the switchbox of the house electricity supply, replacing the conventional bank of fuses.

The principle of the ELCB is that it measures the current on the live side of the protected appliance and the current on the neutral side. These two currents should be exactly equal. However, if current is leaking to Earth at some point, possibly through an unfortunate operator, the two currents are not equal. The ELCB must then shut off the supply as fast as possible. Most ELCBs used in the home are rated to detect a current of 30 mA or less and to switch off the power within 20 ms. More sensitive ECBs used in hospitals and surgeries limit the flow to less than 10 mA.

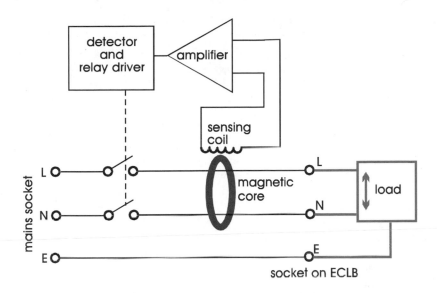

An ELCB opens the switches when the imbalance between currents on the live (L) and neutral (N) sides exceeds 20 mA.

An ELCB is intended to reduce the danger to the operator when there is a fault in the equipment. However, should the operator accidentally provide a direct connection between the live and neutral lines, then no leakage occurs and the device does not cut off the supply.

Measuring high temperatures

Devices for measuring temperature have been described in Chapter 13. Here we look at two techniques for use at temperatures so high that the sensor would be destroyed if placed in the hot environment, such as a furnace. Instead of using a thermometer, we use a **radiation pyrometer** to sense the type and amount of radiation being emitted from a furnace and use this to estimate temperature. Measuring the temperature of very hot objects calls for similar instruments. Pyrometers are based on various sensors, including thermopiles, thermistors and pyroelectric devices, all described in Chapter 11. A pyrometer often has a telescope built into it, so that it may be directed accurately at the open furnace door or the hot object. This may incorporate a dichroic mirror to allow visible light to pass through to the telescope but to reflect infra-red on to the sensor.

Some pyrometers work on the **chopped-beam** principle. A bladed 'fan' in the optical system alternately exposes the sensor to radiation from the furnace and to radiation from the filament of a standard lamp. The current through the filament is adjusted automatically until the sensor receives equal amounts of radiation from either source. In other words the output from the sensor is steady, not alternating. The temperature of the furnace is then found by referring to calibration charts.

A **bolometer** does not sense infra-red directly, but measures its heating effect on a thermistor. The instrument has a matched pair of thermistors, one of which is exposed to the radiation, the other of which is shielded. The exposed thermistor is blackened to assist the absorption of infra-red. The purpose of the shielded thermistor is to compensate for changes in ambient temperature. The resistances of the thermistors are compared using a bridge circuit. The output of the bridge circuit is proportional to the amount of radiation falling on the exposed thermistor, and hence to the temperature of the hot object.

Moisture content

Very often, as in the example of the gas flow sensor mentioned earlier, it is easy to use a basic electronic quantity for measuring what may be a quite complex attribute of a product. An example of this is using capacitance to measure the moisture content of fabric. The fabric is passed over rollers between two metal plates which act as the plates of a capacitor.

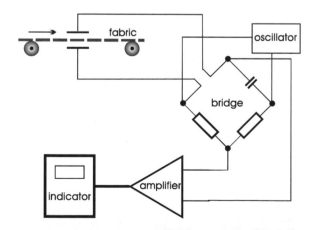

The capacitance between the two plates is measured by a capacitance bridge.

The fabric is thus the equivalent of the dielectric of the capacitor. The capacitance is proportional to the moisture content of the fabric. This may be measured using a bridge circuit. If the capacitance is found to be outside acceptable limits, the circuit may operate an alarm or may automatically take an action such as raising the temperature of driers.

In agriculture, a similar method is used in grain driers. A large specially shaped capacitor is situated in the ducts in which grain is passed through the drier. Changes in capacitance produce signals which are used to control the rate at which the grain passes through the drier, to ensure that it is evenly dried.

Moisture content can also be measured using another electrical quantity, resistance. Sawn timber to be used for making furniture must first be brought to a suitable moisture content, which is 12%. It is dried in a kiln and while it is drying the moisture content of a sample of pieces is measured. Two probes are driven into the sample pieces, a given distance apart. Measurement of the resistance between the probes allows the moisture content to be calculated. This is another illustration of how computers are used to process data obtained electronically. During the kiln drying, which may take up to 4 weeks, depending on the initial value, the moisture content is measured at regular intervals. The data obtained is used to predict when the content will have fallen to 12%. This information is of great use in stock management.

Body scanning

One of the most dramatic advances in medical electronics has been in the field of body scanning. The EMI **whole body scanner** consists of a source of a narrow beam of X-rays and an X-ray detector that are located opposite each other and able to move round a vertical circular path. The body of the patient lies horizontally inside the circle. The detector measures what fraction of the X rays has been absorbed while passing through the body. This information is fed to a computer and stored in its memory. As the apparatus rotates and gradually scans along the body, the 'slice' of the body that is being scanned is penetrated by X-rays from all directions. Parts that are in the shadow of bone when the beam comes from one direction are clear of shadows when the beam comes from another direction. The computer is programmed to process the stored information and produce a display that shows the shapes and positions of the structures present in the slice. Early versions of the EMI scanners were small and used only for scanning the brain. They can locate brain cysts and tumours in scans lasting only a few minutes. More recent equipment can scan the whole body.

Another technique for body scanning is **magnetic resonance imaging** (MRI). Scans of slices though parts of the body, especially the brain, are made by detecting the resonance of atoms in a strong magnetic field. This allows maps to be made according to one of a number of tissue characteristics, such as proton density. The technique requires a very strong and uniform magnetic field throughout the part being scanned. Electromagnets employing superconductors are excellent for producing the fields required. Protons in water and tissues align themselves in the field. A burst of radio-frequency radiation is used to excite the protons, which then release energy as they return to their unexcited state. This energy is detected and the measurements fed to a computer which uses the information to build images of body structure. MRI has an advantage over other scanning techniques that it does not expose the subject to ionizing radiation, such as X-rays or emissions from radioactive substances. As in several previous examples, the measuring is electronic but the subsequent processing of the data depends on powerful computers.

A further application of MRI is **projection angiography.** This relies on the fact that the faster the flow of blood, the brighter the image on the MRI screen. This technique is used to measure rates of blood flow in regions of the body such as the head and neck. Information gained in this way is useful in the treatment of strokes and cerebrovascular diseases.

Measuring bodily potentials

Electronic equipment is used to measure the small electrical potentials developed in the body as a result of the activity of muscle and brain tissues. By monitoring these potentials it is possible to detect abnormalities in bodily functions and so diagnose diseases. The heart is a powerful muscle and as it beats it generates potentials which spread throughout the body. The physician can measure and record these potentials by means of electrodes connected to the patient's limbs. The potentials conform to a pattern that is the same in all normal people so that divergence from the pattern indicates certain types of abnormality. This technique is known as **electrocardiograph** (or ECG for short). The potentials generated by the heart are picked up by silver or platinum electrodes attached to the wrists and a leg of the patient. The electrodes are connected to the input terminals of sensitive instrumentation op amps. Measurements are made in turn of the variations in potential difference between the right arm and left arm, the left arm and the leg, and the right arm and the leg. Because the body is a high-impedance source (that is, the potentials are measurable but are unable to deliver a large current to the apparatus) the amplifier has FET inputs, which have high impedance

The output of the amplifier is processed and a permanent record of the signals from the patient is printed out.

Potentials in the brain are explored with similar equipment to produce an **electroencephalograph**. High-gain amplifiers are needed as the brain is less physically active than the heart and produces potentials of only about 100 microvolts. The potentials are detected in the patient's scalp by using small disc electrodes firmly attached to the skin by a harness. It is usual to have up to 24 electrodes. Each provides a series of signals which are fed to separate amplifiers. The signals can be displayed simultaneously on a computer monitor or printed out.

Radio pill

This is an example of remote measurement. Radio pills are, as the name suggests, devices to be swallowed. The idea is that they sense the conditions in different parts of the alimentary canal as they pass through them. The information is transmitted by the pill as a radio signal which can be picked up and analysed by the physician while the pill is passing through the body. Before the advent of the transistor, it was impossible to make a device small enough to be swallowed without considerable discomfort to the patient. Using a transistor powered by a miniature cell, a pill need be only about 20 mm long and 8 mm in diameter. A typical radio pill senses and transmits information about the pressure, temperature, acidity and several other factors of interest to the physician. The pill is essentially a minute radio transmitter working at about 300–400 kHz. The information is radiated through the body to a receiver outside, which detects the signals and passes them to a recorder. Information is conveyed by varying the frequency of the carrier wave and this can be done in various ways. If, for example, we want to know the pressure inside the intestines, an inductor in the circuit is made to alter the tuning of the transmitter, its iron core being displaced in accordance with the pressure on the walls of the pill. If the acidity inside the stomach is to be measured, this can be done by placing pH electrodes on the surface of the pill. When immersed in the acid solution in the stomach, the electrodes generate a p.d. proportional to the acidity. This is used to alter the frequency of the transmitter by altering the potential of the base of a transistor. If temperature is to be measured, the effect of temperature on the characteristics of the transistor can be used to vary the frequency.

26 Electronic control

Electronic techniques have made possible great advances in the control of a wide range of systems, from dishwashers to nuclear power stations. The control that we describe in this chapter is essentially automatic. All the operator has to do is to make a few settings and press the button. Electronics does the rest. In manufacturing and some other industries this kind of control is called **automation** but the same principles are found in all manner of control systems from steering a radio telescope to manoeuvring a Moon-buggy.

A simple instance of a basic automatic control system is a **thermostat** circuit (opposite). The thermostat controls a heater so as to keep the room temperature constant. It has three main sections, the **sensor** (thermistor), the **actuator** (relay) and the **processing** stage (op amp). The operation of this circuit relies on the climate being cool, so that room temperature can be kept at a fixed level simply by switching the heater on or off. In a warm climate, we would need to provide some form of cooling too, which would make the system more complicated.

The thermistor (R1) is part of a potential divider. As temperature increases, the resistance of R1 decreases and the potential at junction A falls. The operator sets the circuit by adjusting the variable resistor VR1. VR1 is another potential divider.

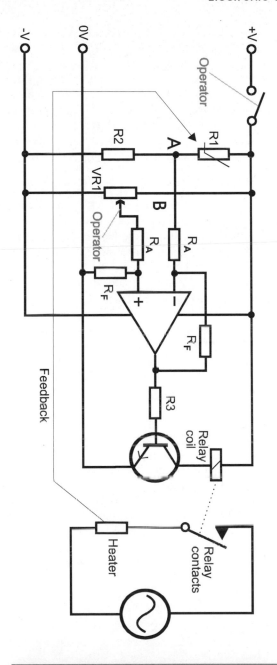

In this thermostat, the op amp compares the potential at A (actual temperature) with that at B (set point)

The temperature at which the room is to be held is known as the **set point**. The higher the set point, the higher the potential at B. The op amp compares the potentials at A and B. If A is lower than B, it means that the temperature of the room is not as high as the set point. The output of the op amp swings high, turning on the transistor. This energizes the relay, its contacts close, and the heater is switched on. However, if A is higher than B, the room has reached the set point, the op amp output swings low, turning the transistor off. This de-energizes the relay, its contacts open, and the heater is turned off. Because of the amplifying action of the op amp, the circuit is sensitive to very small changes of temperature.

The op amp provides one of the essential features of this circuit (it is wired as a comparator). It compares the actual temperature of the room with the set point. If the room temperature is too low, it produces a high output to turn on the heater. This output is known as the **actuator signal**.

Another essential feature of this circuit is that there is **feedback**. Information about the output of the system (the actual temperature) is fed back to the input of the system (the sensor). This happens because air in contact with the heater eventually circulates to the thermistor. The thermistor is wired into the circuit in such a place that, as the temperature exceeds the set point, the heater is turned off. As the temperature falls below the set point, the heater is turned on. The feedback makes the thermostat switch the heater so that it *opposes* changes in room temperature. We call this **negative feedback**.

This type of electronic control is widely used. As another example, suppose we wish a conveyor belt to travel at a certain speed. In this case the actuator is the electric motor that drives the belt. To this is linked a device such as a Hall-effect tachometer (p. 123), the output of this is proportional to the speed of the motor. This signal (which could be either an analogue signal or a digital one) is the feedback of the system.

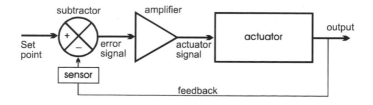

In this diagram, the subtractive and amplifying actions of the op amp are shown separately.

The feedback, detected by the sensor is compared with the set point, using a comparator circuit (perhaps an op amp). In the case of an op amp, the first stage of the op amp subtracts the feedback signal from the set point signal. The difference, which is the **error signal,** is then amplified in the next stage of the op amp.The set point is set manually or by computer to make the motor run at the correct speed. The error signal, which depends on the difference between the actual speed of the motor and its correct speed is amplified and then used to control the power delivered to the motor. In this way the speed is automatically regulated. There is some slight error in this type of system because deviations from the set point can not be corrected until after they have occurred, but this error can be kept very small by careful design of the system.

Feedback control can be devised for a wide range of quantities including humidity, flow, pressure and weight. Circuits may be analogue or digital without affecting the principles involved. A Gray-code sensor may be used to provide feedback in a digital circuit.

As its name implies, a thermostat is intended to hold temperature at a fixed level. The same applies to the example of the conveyor belt. In other words, the circuit is a **regulator**. However, if the set point is programmed to change at different stages in an operation, the actuator can be made to perform a sequence of operations automatically. For example, a tank of dye solution can be brought to different temperatures at different stages in the processing. The power steering system of a car is another example. Systems like these, in which the set point is frequently or even continuously changed and the actuator responds accordingly, are known as **servo systems.**

With appropriate software, a computer can take over many of the functions of a control system, except for the sensors and actuators. An example of this is found in machine tools. Standard control languages have been devised for such operations. For example printed circuit boards are drilled by numerically controlled drilling machines. Using the standard format, the software specifies the size and position of all the holes in a board and the machine drills the holes automatically.

Other responses

The response of the systems described above is binary. The actuator is activated or it is not. These are called **on-off systems** or sometimes **bang-bang systems**. This type of system is the simplest to design, the cheapest to install, and the easiest to operate and maintain. For these reasons, these are by far the commonest type of system.

Some applications need a more quantitative response. For example, a gantry robot has to lift a package and transport it a distance before setting it down. The aim is to perform this task as quickly as possible. If the robot moves too fast, it may be difficult to put the load down in exactly the right place. If it moves too slowly, it takes an unnecessarily long time to complete the task. We need a system that varies the speed according to how far the gantry has to move.

Speed can be a maximum when the gantry is a long distance from the setting-down point. However, it must move more slowly as it gets closer to this point. This can be done by **proportional control.** The error signal does not simply switch the actuator on or off. Instead, the response of the system (the speed of the motor, for example) is proportional to the *size* of the error signal. As the gantry approaches the set-down point the error signal decreases and the speed of the motor decreases.

A gantry robot is used to stack heavy blocks of printed paper.

Proportional control is an elementary example of the sophisticated control techniques available to the engineer. Systems like this are built from op amps, or can be controlled by a computer. However, it may be more economic to ignore the finer points of proportional and other forms of control and to employ an on-off system. Take as an example, a factory filling cartridges with toner powder for photocopiers. It is possible to devise a system to control precisely the amount of powder loaded into each cartridge as it passes along the production line. However, it is cheaper to fill the cartridges by a less refined method and, at the end of the process, to simply reject those that are over- or under-weight. In industrial practice, simplicity and low cost outweigh elegant control theory.

Logical control

On page 219 we showed how electronic logic gates can be used to control the switching of lamps. The system has the same three divisions as the analogue systems described above — sensors, processing and actuators. In this case, the processing is by logic ICs. Once the requirements of the system have been clearly listed, the logic circuit is designed and built, using standard logic ICs. Such a system will work well for many years. However, it may subsequently be found that it does not work perfectly as had first been supposed, or perhaps the requirements may change, or the system may need to be expanded. Altering a logic system often means redesigning parts of the circuit board or adding new boards. It is often preferable to scrap the existing system and start again. This is where computer control has the advantage, either with a PC or with a microcontroller. If the system needs amending, expanding or updating, the biggest change is in the software, not the hardware. Additional or different input or output ports may be needed, but there is no logic board to be redesigned, re-tested and replaced.

Nowadays, computers and microcontrollers are taking over the functions of most logic ICs in control circuitry. An illustration of this is provided by the title photograph of this chapter. This is part of the control room at Ironbridge Power Station, Shropshire. Computer monitors display diagrams of the generating plant, including relevant data for each section. The regulation of each stage is under the automatic control of the computer system. Instead of the meters that would be found on a conventional control panel, the monitors show 'virtual meters' to provide the required readouts. Instead of opening or closing switches or adjusting variable resistors manually, the engineer uses a computer mouse to select or drag the 'virtual controls'. The whole power station is operated using a mouse.

PLCs

A **programmable logic controller** is a simple computer-like device especially designed for industrial control systems. PLCs are often used for other purposes too, such as controlling the automatic microwave systems that open and close shop doors. A typical PLC is based on a specialized microprocessor. There is a terminal which usually has a few keys for operating the PLC and an LED or LCD display to show the current status of the system. The terminal also contains the program and data memory. The control sequences and logic are set up on a PC or a special programming unit which is not part of the PLC. When the program has been written it is downloaded into the memory of the PLC, which then runs independently.

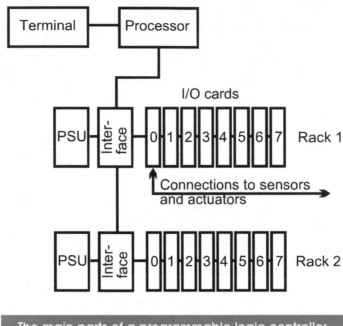

The most noticeable feature of a PLC is the large number of inputs and outputs. Each input and output 'card' communicates with the outside world through a number of ports, usually 8 or 16 ports for each card. There may be several racks of cards, because controlling a large industrial unit such as a production line or a printing press involves numerous sensory inputs and a corresponding number of outputs. However, for elementary tasks there are smaller PLCs that have only about 6 inputs and outputs, each complete with interface circuits and housed as a single unit.

The idea behind PLCs is that they comprise a number of ready-made processing and interfacing devices that may be purchased 'off the shelf' for assembling into a wide variety of different systems. The system is easily expandable and reconfigurable as requirements change. Several types of interface cards are available, for input and output, for analogue or digital signals, and for AC or DC. All have in common the ability to isolate the PLC from the heavy currents, large voltage spikes and general electrical noise that is all too common in industrial plant. In the example opposite, the input circuit is separated by an optoisolator from the circuit that forwards the input data to the processor.

The door of this PLC cabinet is open to reveal three PLCs on the right, with numerous connections to the sensors and actuators of the colour-printing press that they control. Relays on the left are used for switching.

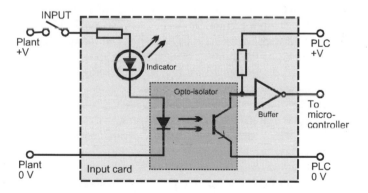

This is one of the 8 or more channels of a typical DC digital input card. The supply from the plant may be at a relatively high voltage, such as 24 V. That used by the PLC is often at 3 V or 5 V and is regulated.

Although most PLCs are capable of high-level functions, they are more often used in situations that require only elementary processing. In such instances, they do not need a computer expert to program them. PLCs use special programming languages that are easily learned by the engineers on the site. One such PLC, known as SPLat, can be programmed for most simple tasks using a BASIC-type language with only 11 different instructions in its set.

Electronic control systems relieve human operators of the tedious tasks of measuring, assessing and correcting equipment and, moreover, can operate at much faster speeds. They also have the advantage that they can work for long hours without tiring. They do not forget to perform each step in their task. As explained in the next section, many factory jobs such as painting, welding and assembling of parts can be performed exclusively by electronically controlled robots. The social effects of this have been notable, bringing unemployment to some and new careers to others. One of the lesser publicized consequences of automation is that large factories can no longer support the canteen where dozens or hundreds of workers formerly took their meals. Now the canteen is shut down and the few remaining operatives eat in the local café.

Robots

Many people's idea of a robot is something like C-3PO, the humanoid creature featured in the film *Star Wars*. In reality, a typical robot is of the **articulated arm** type shown in the photograph opposite. This has a fixed base and an arm consisting of a number of segments. The segments can rotate in six axes, giving the arm a high degree of flexibility. The arm usually ends in a special tool such as a paint spray or a welding torch (as in the photograph), which can be positioned with a precision of less than a millimetre.

There are also **moving platform robots**, or **rovers,** that automatically find their way on the factory floor, delivering parts to the production line and receiving completed units. They can also be used for operations in hazardous areas, such as in chemical plant, or on the surface of Mars, or for the secure transportation of drugs in hospitals. Often rovers have 3 or 4 wheels, but wheels cannot negotiate rough surfaces. For such terrain a robot may have a number of legs, but with these comes the problem of maintaining balance. In spite of the problems, some very successful moving platform robots have been designed. These include the underwater rovers that can explore wrecks or repair the underwater supports of oil drilling platforms. They can reach depths of several thousand metres and relay video pictures to engineeers on the surface.

This Hitachi MR6100 articulated arm robot is programmed for welding. It rotates around six axes, as shown by the arrows.

The operation of a robot is controlled by a computer, which can be 'taught' the jobs the computer has to do. The robot is taken through the required routines by a human operator. After one such 'training' session the robot 'remembers' the movements required (for example, to spray-paint the body of a car) and repeats the action on every subsequent occasion without the need for human guidance. The essential feature of robots is that they are flexible; they can be programmed to perform a variety of different tasks, so that the same robot can be used in different parts of the production line, or can be re-programmed when the product design is altered.

At this point in our account of electronic control systems we are going beyond the province of electronics. The action of robots is controlled by software, and their seeming skill is the result of the skill of the programmer. Their reliability is the result of the skill of the engineer. These are aspects which it would be out of place to discuss here, but the electronic sensors used by the robots are very similar to those described in earlier chapters in this book. The electronic circuits of the robot are based on the transistors and logic gates we have already described. Those who have read this far in the book will find nothing new in the electronic aspects of these apparently human machines.

27 Electronics and the future

This chapter is the one that needs more revision than any other when the time comes for a new edition of the book. Electronics changes rapidly and this chapter reflects the changes. Developments which were new at the time of the previous edition may have come to fruition and are now transferred to the earlier chapters of the book. In the 1995 edition we wrote about what was then referred to as the 'Information Superhighway'. Now the Internet is firmly part of everyday life. Today is the era of 'dot-com'. On-line banking, shopping, news reports, tutorials, MP3, and a mine of information are now available on the home computer, usually within a few seconds. Working from home is becoming more of a practical proposition for many people. The Internet facilities promised in the previous edition have all come about. Fuzzy logic and neural networks, mentioned in this chapter in the previous editon, have also become well established and are used in a variety of domestic and industrial equipment.

The trend toward miniaturization has continued too. The range of SMT components expands daily. Mobile telephones and video cameras shrink steadily in size. The newly released *Matchbox PC* is claimed to be the smallest PC in the World. Measuring only 70 mm × 50 mm × 24 mm, and weighing only 93 g, it performs most of the functions of its larger brethren, including running the Windows and Linux operating systems.

But, as always, some developments that were heralded as major break-throughs, subsequently fail to live up to expectation or are superseded by later developments. Quadraphonic sound and bubble memories are two examples.

Another prediction, originating decades ago, has never been fulfilled. This was the belief that the advent of computing on a wide scale would lead to the 'paperless office'. Observation of the desks of most present-day offices shows that, far from reducing the amount of paper consumed, computing has increased the demand for paper. A laser-jet printer can silently gush out reams of beautifully printed paper with little effort. It is hard to explain the lack of enthusiasm for working solely with images on the monitor screen. 'Hard copy' has a nice safe feel to it. One can never quite trust the hard disk not to crash and lose all the data stored on it. There are psychological aspects too. It seems much easier to read and understand a document that is printed on paper. When we were writing this book, most of the work was done directly on the computer. However, it was printed out for close editing in the final stages. The mis-spellings and other typographical errors seem to show up much more clearly on paper than on the monitor screen. And a paper copy is just the thing to scribble on, crossing out errors and adding amendments.

Having partly explained away the failure of the paperless office, we might note a development that could help to bring it about. **Electronic ink** has recently been demonstrated as a practical proposition.

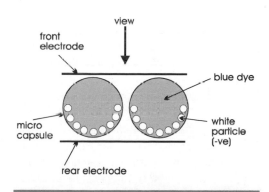

Electronic paper has a layer of millions of microcapsules. They appear to be black or white depending on the direction of the field betwen the electrodes.

A layer of micro-capsules is sandwiched between two flexible, and transparent electrodes. Each capsule contains a blue liquid dye in which are suspended fine white particles of titanium oxide. The particles are negatively charged so that, when the front electrode is made negative and the rear electrode is made positive, the particles are attracted to the rear of each microcapsule. When viewed from above, the capsule looks black.

If the charges on the electrodes are reversed, the particles move to the front of the capsule, which then looks white. The charges are applied by an array of flexible transistors, one for each pixel.

The ink together with the transistor layer makes up what is called **electronic paper**. Text and illustrations are 'printed' on the paper electronically and appear as a black and white image of high contrast, like a page printed in conventional ink. The paper has a wide angle of view, unlike many LCD displays. The image remains after the power is switched off, but can be cleared and replaced by a new image. This technology is only in its infancy and the aim is to develop it to the stage of being able to produce electronic books. Perhaps by the time of the next revision of *Understand Electronics*, electronic paper and ink will be familiar in our paperless offices.

Electronic ink and most other innovations in electronics are related to increasing computing power. The **Blue Gene** programme undertaken by IBM is aimed at producing a new computer architecture that will comfortably out-perform all of today's machines. The new architecture is called **SMASH**, which is an acronym for **S**imple, **MA**ny and **S**elf-**H**ealing. The system is simple because it is based on RISC (reduced instruction set) processors. These processors have only 57 instructions compared with several hundred in a Pentium. The term *many*, refers to there being as many as 30 processors on one chip, all working in parallel. The chips will be mounted on boards, each carrying 36 chips. The whole computer will have 5 stacked arrays, each of 16 x 16 boards. With so many processors all working at once, the Blue Gene will be about two million times more powerful than the best of today's PCs. However, such a large system will inevitably be subject to failure of some of its parts at fairly frequent intervals. This can be overcome by incorporating self-checking routines. With serious failures, parts of the system can be shut off and the computation will proceed with the remaining sound processors. This is the self-healing aspect of the computer's architecture.

Such are the foreseeable developments in computer architecture, but what of future developments in electronics itself? Here the most likely forecast can be summed up in three words: smaller, faster, and smarter. The more significant aspect of size reduction is the miniaturization of the components themselves, to which we have already referred. Much research is aimed at making transistors smaller than ever, with the aim of packing more and more of them on the chip. New semiconductors, such as those based on gallium arsenide, promise to give us smaller, more densely packed transistors, leading to a new phase of larger-scale integration. Small size is not the sole aim of these researches; a densely-packed circuit has shorter connections between its components, which means that signals pass from one to the other more quickly.

Operating speed can be significantly increased. A parallel effect arises because smaller transistors have lower capacitance and switch states more rapidly. Power consumption is reduced too. All of this encourages the use of microprocessors and other complex ICs in electronic systems, leading to an overall increase in 'smartness' of all our electronic equipment. This trend is set to continue, and our smaller-sized circuits also will be both faster and smarter. The three qualities go hand-in-hand.

Apart from enhancing the performance of semiconductors and looking for new ones, researchers are examining other ways of building components and circuits based on new technologies. At present, the semiconductors in use in electronics are elements (silicon, germanium) or inorganic compounds (cadmium sulphide, gallium arsenide). Now the semiconducting properties of some of the organic compounds are under investigation. For example, it has been shown that a film of phenylene vinylene emits a greenish yellow light when a p.d. of 1.5 V is applied across it. This organic LED has very low efficiency, but research is showing ways of increasing this. Advances in this field will no doubt lead to the discovery of new organic semiconductors and eventually to a range of components based on this new technology.

Another branch of electronic research seeks to produce the smallest possible components, consisting of single molecules. The behaviour of certain kinds of molecules is analogous to the behaviour of semiconductor materials. It is feasible that we could design and build molecules that will perform the same functions as transistors and other semiconducting devices. **Molecular electronics**, as it is called, would allow circuits to be made very much smaller than is possible with techniques for fabricating integrated circuits from semiconductors. One of the biggest problems will be connecting such one-molecule devices to external circuitry. This difficulty is one of the more important impediments to the development of molecular devices.

One line of research is aimed at producing **chemically assembled electronic nanocomputers**, or CAENs. Already an electronic switch has been built from a single molecule of rotaxane. This y-shaped molecule has a ring-shaped band of electrons on one of its arms. The band can be positioned either at the end of one arm or further down the arm. The molecule is located at a crossing point in a lattice of two sets of parallel wires. The arrangement is similar to that of the transistors in the drawing on page 218. When the electrons are at the end of the arm, current can flow from a wire in one set of conductors (at positive potential) to a wire in the other set (at ground potential). No current can flow when the electrons are further down the arm. Thus we have a molecular switch, which one day may be used as the basis of a microprocessor the size of a grain of salt.

This is some way off yet, but plans are in hand to build a circuit capable of simple mathematical calculations. The circuit comprises five logic gates but its dimensions are only 0.2 μm.

If future computers are going to be able to process such enormous quantities of data, they will need correspondingly capacious memories. IBM are working on a new form of data storage, known as **Millipede.** Like CD recording, this is a mechanical technique in which data is stored as an array of pits in a film of polymethylmethacrylate (PMMA). This is coated on to a flat plate of silicon. Conductive probes in a 32 x 32 array are scanned across the plate. When storing data, some of the probes are heated by passing a current through them so that they penetrate the PMMA film to the substrate, forming pits. Data is read by slightly warming the probes so that those contacting the substrate lose heat. The heat loss is measured and indicates whether the bit stored at each point is a '1' or a '0'.

At present the density of magnetic data storage on a hard disk is just over 3 gigabytes per square centimetre. It seems that, with projected improvements, the top limit for hard disks is about 15 GB/cm^2. With the Millipede technique, the density is in the order of 80 GB/cm^2.

Reduction in the size of components reached a new record in December 2000 when Intel released the news that it had built the World's smallest and fastest CMOS transistor. The device measures only 30 nm across and 60 nm high (1 nanometre is one thousandth of a millionth of a metre). To obtain the high definition for printing the masks for this device, Intel used Extreme Ultraviolet Lithography. This success opens the way to building a microprocessor with 400 million transistors running at a speed of 10 GHz, and operating on a supply of less than 1 V. This will considerably reduce power requirements. However, there are still problems to overcome, particularly in reducing the 'on' resistance of the transistor. Intel has suggested that, once these problems are conquered, we may well see a processor capable of driving a universal translator, to convert text from one language into a wide range of other languages.

Electron electronics is the name given to the technology of devices that operate using single electrons. These could be the smallest and the fastest of all future components. Not only would they be fast, but they would permit the highest possible density of data storage. So far, IBM has developed a single electron storage of one bit of data. The device consists of an ellipse of single cobalt atoms measuring about 20 nm × 10 nm arranged on a substrate. It is not possible here to go into the way it works. Indeed, there is still a lot for researchers to find out about this technology.

There will also be many problems to overcome — how to connect the single-electron devices together, and how to interface them to the outside world. We hope that the next revision of this chapter will carry news of progress.

In the decades following the invention of transistors there was a period of exploitation of their properties. In the nineteen-seventies, new designs of integrated circuits appeared almost daily. Within a few years, the manufacturers had produced ICs to undertake almost all of the special functions that we could ever need. More recently, there have been fewer really novel ICs. Perhaps the most unusual have been the processors designed for implementing fuzzy logic and neural networks. But most of the new designs have been simply improvements on the old ones. Faster action, lower power consumption, and enhanced ability to work in hostile environments have been the main directions of improvement.

Now that microprocessors and microcontrollers are so versatile and cheap, the design and testing of a new system is easy. If there is a function to be performed and there is no IC ready-made to perform it, take a general-purpose microcontroller, equip it with the necessary electronic sensors and output devices, then program it to perform that function. Once this stratum of electronic circuit design has been reached, there is less need for new circuit designs and new ICs.

Software is taking over many of the functions formerly fulfilled by complex electronic circuits. For example, instead of using capacitors and inductors to build an audio filter, we can perform the same task by letting a computer or microcontroller operate in real time on the digitized audio signal, and with superior performance.

From the above, and from the accounts of research in progress, it seems that the principal contribution of electronics in the foreseeable future will be to provide the means for building up massive computing power. But, whatever the future holds, it is clear that electronics will continue to play a major part in all of our lives.

Acknowledgements

The author thanks the companies and organizations concerned listed below for permission to take photographs on their sites:

Bossong Engineering Pty. Ltd., Canning vale, Western Australia (p. 321).
Eastern Generation Ltd., Ironbridge, Shropshire (p. 312).
Jarrold Printing Ltd., Norwich (pp. 316 and 319).
Kronospan Ltd., Chirk (p. 297).
Nuffield Radio Astronomy Laboratories, Jodrell Bank, Cheshire (p. 182).

The author also wishes to thank the following individuals for helpful advice in explaining how they make use of electronic control systems:

R. Byrnes (Richard Burbidge Decorative Timber Ltd., Oswestry, Shropshire), Russ Cason (Senior Electrical Engineer, Kronospan Ltd., Chirk), Phil Ludgate (Parts Engineering Manager, Ricoh UK Products Ltd., Telford), Ian Morison (NRAL, Jodrell Bank), Dave Potter (Head of Process Control Section, Eastern Generation Ltd.), Philip Ringwood (Electronics Engineer, Jarrold Printing Ltd.), Clayton Steele (Automation Technical Manager, Bossong Engineering Pty. Ltd.), Nick Thompson (Kronospan Ltd., Chirk).

Thanks also to Nic Bishop for on-the-spot information about the SETI project at Arecibo Radio Observatory, Puerto Rico.

Index